The Breeding of Animals

by F.B. Mumford

with an introduction by Jackson Chambers

This work contains material that was originally published in 1921.

This publication is within the Public Domain.

This edition is reprinted for educational purposes and in accordance with all applicable Federal Laws.

Introduction Copyright 2016 by Jackson Chambers

Self Reliance Books

Get more historic titles on animal and stock breeding, gardening and old fashioned skills by visiting us at:

http://selfreliancebooks.blogspot.com/

Introduction

I am pleased to present yet another title on the Principles of Animal Breeding.

This volume is entitled "The Breeding of Animals" and was published in 1921.

The work is in the Public Domain and is re-printed here in accordance with Federal Laws.

As with all reprinted books of this age that are intended to perfectly reproduce the original edition, considerable pains and effort had to be undertaken to correct fading and sometimes outright damage to existing proofs of this title. At times, this task is quite monumental, requiring an almost total "rebuilding" of some pages from digital proofs of multiple copies. Despite this, imperfections still sometimes exist in the final proof and may detract from the visual appearance of the text.

I hope you enjoy reading this book as much as I enjoyed making it available to readers again.

Jackson Chambers

PREFACE

THE problems of the animal-breeder may all be grouped under the three subjects, reproduction, inheritance, development. The mere multiplication of the species is now, and always has been, the major work of the breeder of domestic animals. But the real breeder is not only concerned with the production of mere numbers of animals of a given species but is primarily interested in securing animals possessing the largest number of desirable qualities and the least number of qualities undesirable to man.

How to maintain the good qualities that have already appeared in an individual, and how to cause other and better qualities to become dominant in future individuals of the same species, is the problem of inheritance that chiefly concerns the breeder of the domestic animals. The highest attainments in the breeder's art have come only to those who have had a good knowledge of the principles and laws of heredity. The development of animals from the fertilization of the egg to maturity and their proper maintenance throughout their productive lives is second in importance only to inheritance. The environment of the animal, including food, climate, and exercise of functions, determines the degree of development in the individual animal. The term Development, as used in this connection, has reference to the unfolding of capabilities that have come to the animal through inheritance.

PREFACE

The author has made no attempt to write a book on genetics or evolution; but the principles of genetics as they apply to the practice of animal-breeding are discussed, in accordance with the conclusions of biologists. The problems of the animal-breeder are in many important particulars widely different from those of the plant-breeder; and the emphasis has been placed on those principles and practices that belong peculiarly to the province of the animal-breeder, while not neglecting the lessons and illustrations to be drawn from the other field.

It has been the purpose to make a practical book which shall be directly useful to the student and to the breeder of animals, and the lessons and examples of which can be applied in the laboratory and on the farm.

F. B. MUMFORD.

COLUMBIA, MO.
November, 1916.

TABLE OF CONTENTS

CHAPTER I

THE CELL 1–15

The cell theory, 1; The germ-cells, 2; The cell, 3; Is the cell the physiological unit? 4; The structure of the cell, 5; Protoplasm, 6; The nucleus, 7; Growth by cell division, 8; How cells divide, 9; Prophase, 10; Metaphase, 11; Anaphase, 12; Telophase, 13; The germ-cells in detail, 14; The ovum, 15; The spermatozoön, 16.

CHAPTER II

REPRODUCTION 16–51

Asexual reproduction, 17; Sexual reproduction, 18; The reproductive process, 19; Oviparous animals, 20; Primary and secondary sexual characters, 21; The reproductive organs of the male, 22; The testicles, 23; Castration, 24; The reproductive organs of the female, 25; The ovaries, 26; The Fallopian tubes, 27; The uterus, 28; The mammary glands, 29; Structure of mammary glands, 30; Fertilization of the ovum, 31; The nature of fertilization, 32; The process of fertilization, 33; The chromosomes, 34; Results of the union of egg and sperm, 35; Changes in the ovum, 36; Changes in the spermatozoön, 37; The significance of reduction, 38; The origin of the germ-cells, 39; Maturation and reduction in the female (oögenesis), 40; Reduction in the male (spermatogenesis), 41; The period of the œstrum or heat, 42; Artificial insemination, 43; Methods of artificial insemination, 44; Conditions influencing the vitality of the sperm-cells, 45; Effect of too

TABLE OF CONTENTS

frequent breeding on the sperm-cells, 46; Vitality of spermatozoa within the female generative organs, 47; Effect of intoxication of the male parent on his offspring, 48; Effect of lead poisoning on the male germ-cells as indicated by the offspring, 49.

CHAPTER III

THE BREEDING SEASON 52–65

Changed conditions, 50; Phases of the breeding season, 51; Procestrum, 52; Œstrum, 53; Metœstrum, 54; Diœstrum, 55; Puberty, 56; Conditions influencing puberty, 57; The œstrum and lactation, 58; Heat during pregnancy, 59; Superfœtation, 60; Examples of superfœtation, 61; Recurrence and duration of the œstrum, 62; Effect of ration on recurrence of œstrum, 63.

CHAPTER IV

GESTATION AND LACTATION 66–84

Gestation: Indications of pregnancy, 64; Physical examination for pregnancy, 65; The period of gestation, 66; Causes of variation in length of gestation period, 67; Incubation, 68; Parturition, 69; Normal parturition of the domestic animals, 70; Mal-presentations, 71; Normal presentations, 72; Treatment for mal-presentation, 73.

Lactation: The mammary glands, 74; The duration of lactation, 75; The food supply, 76; Habit, 77; Heredity, 78; Exercise, 79; Climate, 80; Unusual lactation, 81.

CHAPTER V

FERTILITY 85–112

The number of young at a birth, 82; Period of gestation and fertility, 83; Fertility and the frequency of the recurrence of the œstrum, 84; Fertility and gestation, 85; Duration of the reproductive period, 86; Confinement and fertility, 87; The fertility of domesticated animals, 88; Age and fertility, 89; Relation of age to fertility in

TABLE OF CONTENTS

PAGES

swine, 90; Influence of age of sow on size of litter, 91; Relation of age to fertility in sheep, 92; Influence of age of ram on fertility of ewes, 93; The effect of the age of poultry parents on the offspring, 94; Age and fecundity, 95; Nutrition and fertility, 96; Excessive food supply and nutrition, 97; Other factors affecting fertility, 98; Relation of number of mammæ in swine to fertility, 99; Twins, 100; Characters correlated with fertility, 101; Inbreeding and fertility, 102; Cross-breeding and fertility, 103; Unusual fertility, 104; Unusual fertility among horses, 105; Unusual fertility among cattle, 106; Unusual fertility among sheep, 107; Unusual fertility among swine, 108; Unusual fertility among poultry, 109.

CHAPTER VI

STERILITY 113–130

The causes of sterility, 110; Causes of sterility in the male, 111; Sterility in the female, 112; Closure of the cervix, 113; Obstruction of Fallopian tubes resulting from excessive fatness, 114; Other causes of barrenness, 115; Sterility from fatty degeneration, 116; Sterility caused by abortion, 117; Contagious abortion and sterility, 118; Treatment for contagious abortion, 119; Diagnosis of contagious abortion, 120; The complement fixation test, 121; Sterility of free-martins, 122.

CHAPTER VII

HEREDITY 131–156

Development, 123; Heredity defined, 124; Heredity and variation not antagonistic, 125; The kinds of heredity, 126; Blending inheritance, 127; Alternative inheritance, 128; Particulate or mosaic inheritance, 129; Mendelian inheritance, 130; The experiments of Mendel, 131; The law of dominance, 132; The law of segregation, 133; Unit characters, 134; Gametic purity, 135; Application of Mendel's law, 136; The complexity of animal characters, 137; The inheritance of polled and horned character in cattle, 138; Theory of pure lines,

TABLE OF CONTENTS

139; Hallett's wheat breeding, 140; The presence and absence hypothesis, 141; The theory of mutations, 142; Two important classes of variation, 143; Kinds of mutations, 144; Importance of the mutation theory, 145; Mono-hybrids and di-hybrids, 146.

CHAPTER VIII

INHERITANCE OF ACQUIRED CHARACTERS 157–182

Belief in transmission of acquired characters, 147; Practical breeders believe in transmission of acquired characters, 148; Nature and nurture, 149; What are acquired characters? 150; Somatoplasm and germplasm, 151; Examples of acquired characters, 152; Food supply, 153; Influence of the amount of food on body weight, 154; Food supply and body changes, 155; Influence of limited food supply from birth, 156; Telegony, 157; The Lord Morton mare, 158; The Penycuik experiments, 159; Telegony and mule hybrids, 160; Example of horse foals, 161; Possibility of influence from a previous impregnation, 162; Xenia in animals, 163; Xenia among poultry, 164.

Objections to the Theory that Acquired Characters are Transmitted: No mechanism for the inheritance of acquired characters, 165; The inheritance of disease, 166; Acquired diseases, 167; Congenital disease, 168; Predisposition to disease, 169; Immunity, 170.

CHAPTER IX

HEREDITY AND SEX 183–194

The significance of conjugation and fertilization, 171; Secondary sexual characters, 172; Secondary sexual characters and vigor, 173; Effects of castration and ovariotomy on the secondary sexual characters, 174; Effect of transplanting sexual glands, 175; Effect of internal secretion, 176; Sex-linked characters, 177; Color-blindness, 178; Controlling the sex of offspring, 179; Age or vigor of parents, 180; Comparative vigor or

TABLE OF CONTENTS

sexual superiority, 181; Nutrition and sex, 182; The maturity of the ovum, 183; Seasonal variations in proportion of sexes, 184; Sex cannot be controlled by external conditions, 185.

CHAPTER X

VARIATION 195–216

Importance of variability, 186; Morphological variations, 187; Physiological variations, 188; Meristic variation, 189; Functional variations, 190; Examples of functional variation, 191; Variation in fertility of animals, 192; Variation in the milking function, 193; Variations among different cows, 194; New characters originate in the germ-plasm, 195; Mutilations, 196; The Brown-Sequard experiments, 197; Causes of variation, 198; Cell division a cause of variation, 199; Influence of use and disuse in causing modifications, 200; Importance of causes of variation to the breeder of domestic animals, 201; Germinal variations, 202.

CHAPTER XI

IN-BREEDING 217–242

Definitions, 203; Advantages claimed for in-breeding, 204; Bad results from in-breeding, 205; Decreased fertility and vigor from in-breeding, 206; Darwin's researches, 207; In-breeding cattle, 208; The Chillingham cattle, 209; Deer in parks, 210; In-breeding among pigs, 211; In-breeding sheep, 212; In-breeding dogs, 213; Cornevin's experiments, 214; Weismann's and Von Guaita's experiments, 215; Researches of Ritzema Bos, 216; The Wistar Institute experiments, 217; In-breeding Berkshires by Mr. Gentry, 218; In-breeding corn, 219; How long is it safe to continue in-breeding? 220; Selection important, 221; The truth about in-breeding, 222; Fixing characters by in-breeding, 223; In-breeding and prepotency, 224; Results of in-breeding vary with different species, 225.

TABLE OF CONTENTS

CHAPTER XII

Cross-breeding 243–254

Permanent and temporary results of cross-breeding, 226; Advantages from cross-breeding, 227; Grading, 228; Cross-breeding to increase fertility, 229; Cross-breeding to increase size and restore constitution, 230; Crossing and heredity, 231; First cross and improvement, 232; Cross-breeding as a cause of variation, 233; Crossing species, 234; Crossing bison and cattle, 235; The mule hybrid, 236; The hinny hybrid, 237; Crossing the horse and the zebra, 238; Crossing cattle and zebu, 239; Sheep-goat hybrid, 240.

CHAPTER XIII

Development 255–279

Growth, 241; The growth impulse, 242; Factors influencing growth, 243; Growth and food supply, 244; Capacity to grow, 245; Growth and the cell, 246; When the growth impulse is strongest, 247; Development of the fœtus, 248; Heredity and fœtal development, 249; Birth weight of lambs, 250; Effect of protein and ash in ration on fœtal development, 251; High calcium rations for pregnant swine, 252; Size and vigor of fœtus as influenced by corn and wheat rations, 253; The permanent effect of retarded growth, 254; Early stunting and the capacity to grow, 255; Climate, 256; The age factor in animal-breeding, 257; Premature breeding decreases size, 258; Decreased size due to early breeding not inherited, 259; Influence of early pregnancy on the mother, 260; Gestation and lactation in relation to growth, 261; The Missouri experiments, 262.

CHAPTER XIV

The Practice of Breeding 280–303

Improvement in size, 263; Improvement in function, 264; The milking function, 265; Improvement in wool production, 266; Improvement in tendency to lay on fat,

TABLE OF CONTENTS

267 ; Improvement in speed, 268 ; Selection, 269 ; Natural selection, 270 ; Methodical selection, 271 ; Importance of selection in animal-breeding, 272 ; Aids to selection, 273 ; The real results of selection in the improvement of the domestic animals, 274; Selection within pure lines, 275 ; Vilmorin's pure line wheat-breeding, 276 ; Selection most useful when genetic factors are not pure, 277 ; Pure line theory not opposed to improvement by selection, 278 ; Pedigree, 279 ; Registered breeding animals, 280 ; Registry associations, 281 ; Community breeding, 282 ; Importance of numbers, 283 ; Selecting the best, 284 ; Selecting chance variations, 285 ; The Burbank method, 286 ; The mendelian method, 287.

ILLUSTRATIONS

FIG.		PAGE
1.	Cell division. Prophases. After Wilson	9
2.	Cell division. Later phases. After Wilson	11
3.	The ovarian egg	13
4.	Human spermatozoa	14
5.	Genital organs of boar. After Ellenberger	21
6.	Sections through ovary of rat	25
7.	Genital organs of mare. After Ellenberger	26
8.	Genital organs of cow. After Ellenberger	29
9.	Genital organs of sow. After Ellenberger	30
10.	Genital organs of bitch. After Ellenberger	32
11.	Normal presentation in mare	76
12.	Posterior presentation	77
13.	Abnormal anterior presentation	78
14.	Abnormal posterior presentation	78
15.	Abnormal transverse presentation	79
16.	Diagram illustrating mendelian inheritance	140
17.	Diagram illustrating mendelian inheritance	145

PLATES

PLATE	
I.	Genital organs of mare. After Ellenberger
II.	Superfœtation in mare
III.	Upper: A mare mule that secretes milk
	Lower: A Free-Martin heifer
IV.	Unusual fertility in a cow. Triplet calves
V.	Normal healthy uterus of sow
VI.	Uterus of sterile sow

ILLUSTRATIONS

PLATE
- VII. Upper: Steer fed ration not restricted
  Lower: Steer fed greatly restricted ration
- VIII. Upper: Steer fed ration not restricted
  Lower: Steer fed for normal growth
- IX. Upper: Steer fed generously, at age 120 days
  Lower: Same at age 27 months
- X. Upper: Dam of seven mule foals followed by filly foal
  Lower: Filly foal born after seven mule foals
- XI. Upper: Eleventh foal following ten mule foals
  Lower: Dam of ten mule foals and their filly foal
- XII. Upper: Ninth foal following eight mule foals
  Lower: Foaled after eight mule foals in succession
- XIII. Upper: Twelfth foal following eleven mule foals
  Lower: Mother of eleven mule foals and their horse foal
- XIV. Upper: Close in-breeding of fox terrier
  Lower: Eighth generation of intense in-breeding
- XV. In-bred Berkshire
- XVI. Cross-bred Hereford-Aberdeen Angus steer
- XVII. Upper: Half-blood buffalo (bison) heifer
  Lower: Cross-bred buffalo-cattle bulls
- XVIII. Upper: A five-year-old hinny
  Lower: Sheep-goat hybrid
- XIX. Effect of food supply on development. Side view
- XX. Same. Front view
- XXI. Effect of starvation on capacity to grow
- XXII. Recovery from starvation
- XXIII. Upper: Cow fed corn products
  Lower: Calf from cow fed corn products
- XXIV. Upper: Cow fed wheat products
  Lower: Calf from cow fed wheat products
- XXV. Permanent effect of retarded growth
- XXVI. Grace Briggs at age 18 years
- XXVII. Duchess Skylark Ormsby
- XXVIII. Sophie 19th of Hood Farm
- XXIX. Daughters of the same sire
- XXX. Three generations showing impressive character of original dam
- XXXI. Result of using a pure-bred sire
- XXXII. Result of using a scrub sire

ACKNOWLEDGMENTS

THE author is indebted to R. Pearl for Plate IV, to J. W. Connaway for Plates V and VI, to P. F. Trowbridge for Plates VII, VIII, IX, XXI, XXII, and XXV, to E. A. Trowbridge for Plate XVI, to M. Boyd for Plate XVII, to L. Monsees for Plate XVIII upper, to W. J. Spillman for Plate XVIII lower, to Hart *et al.* for Plates XXIII and XXIV, to C. H. Eckles for Plates XXVI, XXIX, and XXX, and to H. Hackedorn for Plates XXXI and XXXII.

Grateful acknowledgment is made to all these persons for their valuable assistance.

THE BREEDING OF ANIMALS

CHAPTER I

THE CELL

THE greatest modern contribution to the science of animal-breeding was the formulation of the so-called cell theory. This fundamental biological generalization ranks with the evolution theory in importance in many respects; it has given a definite physical basis for inheritance.

1. The cell theory. — All the higher forms of life are made up of cell units, and from these all parts of the body are constructed. Although various in form, all living cells are alike in having within a mass of protoplasm which Huxley called the " physical basis of life." In the simple one-celled forms all functions are found in the one cell, but in the more complex higher forms a physiological division of labor results in the distribution of functions among the cells. " It is to the cell," says Verworn,[1] " that the study of every bodily function sooner or later drives us. In the muscle cell lies the problem of the heart beat and that of muscular contraction; in the gland cell reside the causes of secretion; in the epithelial cell, in the white blood cell, lies the problem of absorption of food, and the secrets of the

---

[1] Wilson, " The Cell," p. 6, 1911.

mind are hidden in the ganglion cell." It is now clearly apparent that the great questions of reproduction, inheritance and development, involving as they do the problems of embryology and evolution, are intricately bound up with the structure and functions of the cell.

2. The germ-cells. — The heritage of the species is contained within the germ-cell. The microscopic egg of the female carries within its minute structure the germ characters of all the maternal ancestors. The germ-cell of the male, the spermatozoön, holds within its exceedingly minute compass the sum total of all the heritable characteristics of the paternal ancestors. All cells are derived from other cells. Virchow's claim made in 1855 that every cell must have been formed by cell division from some previously formed cell, has now been definitely established. Not only does growth and development take place by cell division in the fertilized egg-cell, but the egg-cell itself is directly derived by cell division from an egg-cell of the immediately preceding generation, and so on indefinitely.

3. The cell. — The cell is a mass of protoplasm containing a nucleus, and both nucleus and protoplasm arise through division of the corresponding elements of a preexisting cell.[1]

The word cell is derived from the Greek and means a hollow chamber. The term came into common use before the form and structure of the cell were well understood. The cell-wall which is characteristic of most cells is not an essential part of its structure. It is also

---

[1] Wilson, "The Cell in Development and Inheritance," p. 19, 1906.

Leydig, "Lehrbuch der Histologie," p. 9, 1857.

Schultze, "Arch. Anat. u. Phys.," p. 11, 1861.

true that living cells are never hollow chambers, but are filled in whole or in part by a colorless, viscid, semi-fluid substance, protoplasm. Many cells are simply masses of protoplasm lacking entirely any kind of an enclosing wall. Lying within the protoplasm is a minute body of spherical form which is the cell nucleus. These two, the protoplasm and the nucleus, are of universal occurrence and are the essential components of a living cell.

4. Is the cell the physiological unit? — A study of the form and function of the cell leads to the inevitable conclusion that in a very real sense the cell is the morphological unit of the organism. In its physiological relations to the cells of the body as a whole, however, it is not to be regarded as an independent unit but rather as a localized center of bodily activities. The individual cell in a multicellular body is influenced, sometimes in a marked degree, by the surrounding cells. The most fundamental problem in growth and development of animals is what and how much influence do the body-cells of one group have over the cells of another group. Is there a physiological connection between adjoining cells? Can a group of cells forming a so-called system like the reproductive system in the animal body be influenced by the soma or body-cells? If influenced at all, can such influence have any effect upon the germ plasm in the nucleus of the germ-cell? Is it probable that the germ characters may be changed by these influences so that the offspring of the parent bodies where these influences have been at work, will correspond in any way to the changes occurring in the parent body? In other words, are acquired habits transmitted? It is to be regretted that the microscopic study of the cell has thrown little light upon this fundamental question. As indicated above, the cell in the animal body is in more or less close physiological relation with the other cells of the body, but these relationships and their bearings upon the fundamental facts of biology have not yet been clearly determined.

5. The structure of the cell. — The cell contains a cell-body and the nucleus. The cell-body is all that portion of the protoplasm not contained in the nucleus. The cell in its simplest form is a rounded mass of protoplasm. This type is found generally in one-celled forms and is the characteristic form of the egg-cell of the higher animals. The fact that the form of cells in the higher plants and animals is not always rounded spherical is due to unequal pressure and the movement of the cells comprising the body.

The nucleus is a definite, clearly-marked body existing within the protoplasmic contents of the living cell, and its relation to growth, reproduction and heredity have given it a commanding position in the study of modern biological problems. Other bodies are often found in the cell, such as food granules, products of excretion, fat globules and crystals. None of these plays an active part in the metabolism of the cell and may be regarded as accidental or at least subsidiary to the major rôle played by the protoplasm itself. Another body generally found in the cell is the centrosome which is concerned with the mechanism of cell division. The cell-wall is generally present in the higher forms of plant and animal life and consists of a membrane which is usually lifeless.

6. Protoplasm. — The protoplasm is universally present in every living cell. It is the most fundamentally important life substance. Huxley aptly designated protoplasm as " the physical basis of life." It is not to be regarded as being a definite chemical substance, as its composition changes. It is a viscid, colorless, semifluid material having a higher index of refraction than water, and hence appears brighter. It was called slime by Schleiden. The protoplasm of the cell has a definite structural arrangement appearing as a meshwork, or reticulum, and a ground substance, or cell-sap, filling the intervening spaces. In addition to these two definite substances there are present in the protoplasm minute granules or microsomes which are distributed regularly or irregularly along the lines of the meshwork. While other materials are often found in the protoplasm, the above materials are regarded as the essential elements of primary importance in the activities of the cell.

7. The nucleus. — The nucleus is the center of the constructive activities of the cell. When the nucleus is destroyed, those processes which result in the growth and development of the organism can no longer take place. Only destructive activities are possible in a cell devoid of a nucleus, and these can go forward for only a limited time. "The nucleus is generally regarded," says Wilson, "as a controlling center of cell activity, and hence a primary factor in growth, development and the transmission of specific qualities from cell to cell, and from one generation to another."[1] Growth is the result of cell division, and the impetus for cell division appears to come from the nucleus. The essential fact in cell division is that a portion of the nuclear material of the parent cell shall pass into the new cell. The new cell in its turn becomes a parent cell, and so the process of growth continues. The nucleus is typically spherical and moves freely within the cell. It exhibits two distinct

[1] Wilson, "The Cell in Development and Inheritance," p. 30.

phases which result from the varying degrees of activity present in the nuclear substance. One phase may be designated as the vegetative or quiescent stage and the other the active stage which is characteristic of that period in the development of the nucleus when the many complicated and significant changes occur which result in cell division and in reproduction.

The typical nucleus during the vegetative stage possesses certain distinct structural forms which are concerned in the many important nuclear activities: (1) The nuclear wall which encloses the nucleus and differentiates the nucleus from the cell-body; (2) The reticulum, which is the primary factor in nuclear activities, and appearing as an irregular network. The reticulum in turn comprises two structures, the linin and the chromatin. The latter is undoubtedly the most fundamentally important organic substance concerned with the growth, development and inheritance of plants and animals. It is apparently the only or chief material which is transmitted from the parent cell to the new or daughter cell by division, and from it all nuclear substance may be reformed. The word chromatin is given to this material because it becomes deeply stained upon the addition of certain well-known reagents. The chromatin may appear in the cell in scattered granules, varying in size and form, or in a single deeply staining mass, but more often the arrangement of the chromatin in the nucleus resembles a network which is closely associated with the clearly differentiated linin. (3) The nucleoli are generally but not always present, and their nature and functions are not well understood. By some authorities the nucleoli are the by-products of the activities going on in the nucleus. (4) The ground substance is a fluid filling the chromatin network and is not stained by ordinary reagents.

8. Growth by cell division. — How does growth take place in a living organism? What are the primary factors concerned in the increase in size? What changes in the organism result in the progressive development of the many useful qualities found in the improved types of the domestic animals? How does a microscopic egg-cell develop into a mature and highly organized animal possessing countless cells of many different forms and exercising various and important functions? A still more fundamental problem, if possible, is, how are the qualities of an individual transmitted from parent to offspring? What is the physical basis of heredity and what are the elements concerned in a study of inheritance?

Many of these questions are answered by a study of the cell and specifically by a study of cell division. The entire tissue structure of the animal body arises by repeated division from the germ-cell. The germ-cell itself is the result of the division of a cell which formed a part of the body of the parent. Thus it is that the germ substance carrying the hereditary material is separated from the parent body by cell division. The fertilized egg by continued cell division passes on to every cell in the body a portion of its own substance. The process of cell division must therefore be regarded as a great fundamental fact in the growth and development of plants and animals as well as one of the most significant and primary facts in the transmission of qualities from parent to offspring.

Growth occurs as the result of continued cell division rather than by any material increase in the size of existing cells.

9. How cells divide. — All cells do not divide in the same manner, but the most typical process is known as indirect division or mitosis, and this will be here described. We have seen that the nucleus contains within its minute compass the active material which stimulates the cell to various activities and determines its physiological destiny. If further evidence was needed on this point, it would be found in the remarkable and interesting transformations which take place within the nucleus before and during the process of cell division.

The cell first passes through a vegetative or quiescent stage, and this is followed by a period of activity finally resulting in the formation of two cells from the original parent cell. Various clearly marked stages or phases are distinguishable in this process which have been accurately described by Wilson. The phases observed are for convenience named: (1) prophase, (2) metaphase, (3) anaphase, and (4) telophase.

10. Prophase (Fig. 1). — In the vegetative stage the chromatin of the nucleus exists in the form of a network. Generally during the prophase the chromatin loses its net-like arrangement and assumes the form of a skein like thread known as the spireme. During this stage the spireme thread stains intensely and is fine and closely convoluted. It gradually becomes thicker and the convolutions become more open, giving rise to the "open spireme." Gradually the spireme breaks up into a number of definite straight or curved rods known as chromosomes. It is usual for the wall of the nucleus to disappear during this phase, and the chromosomes then lie naked in the protoplasm of the cell. It is a significant fact that every plant and animal possesses a characteristic number of chromosomes and that this number is always even.

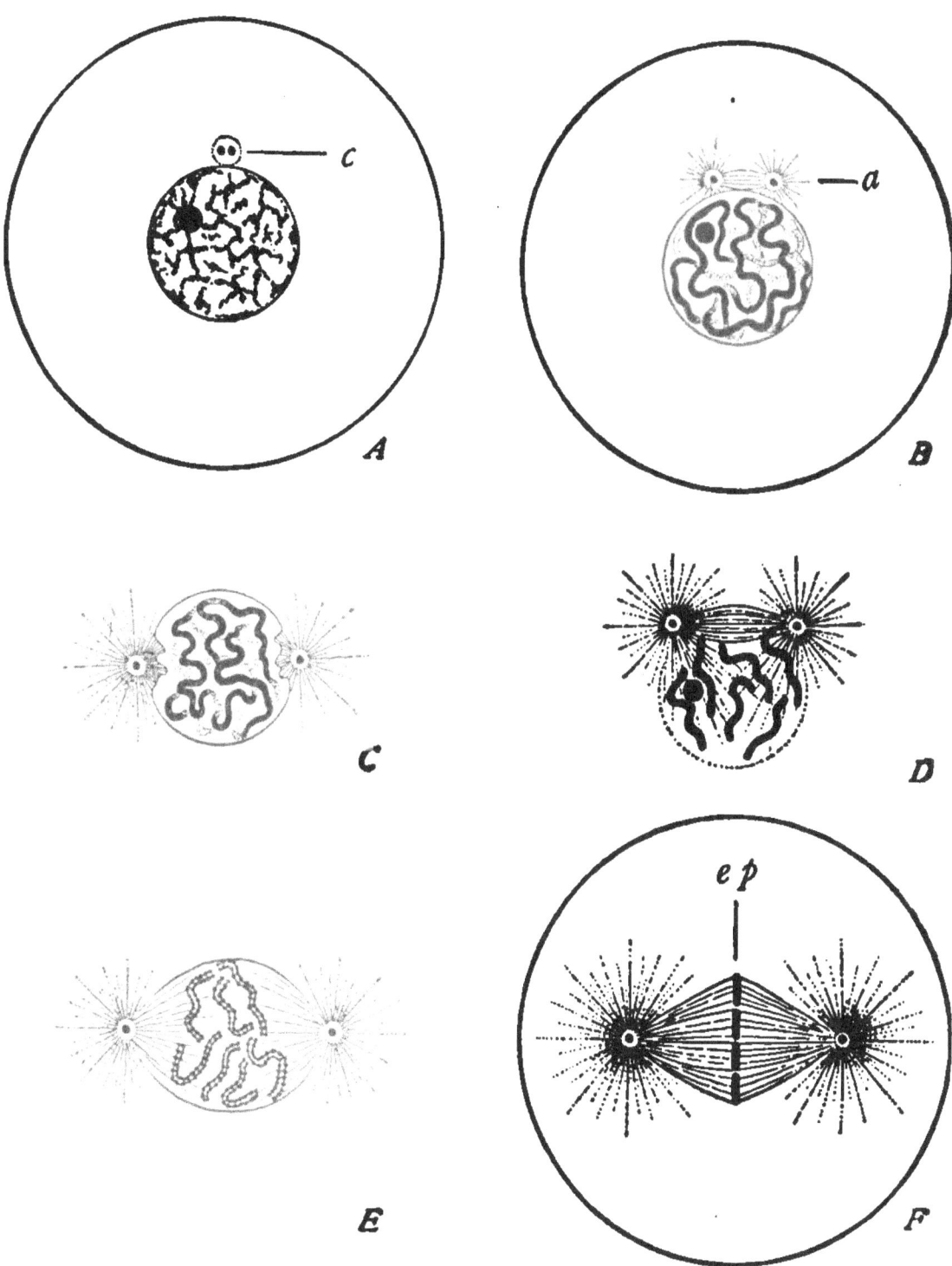

Fig. 1. — Cell division. Diagram showing typical prophases in cell division. *A*, Vegetative or resting stage showing nucleus. *B*, The spireme thread. *C*, Preparation for division. *D*, Formation of chromosomes. *E*, Gradual fading of nuclear membrane. *F*, Chromosomes in equatorial plate ready for division.

In the ox and in man the number is sixteen. The fact that the number of chromosomes is even in all species is due to the fact that during the processes of fertilization one-half the chromosomes are derived from the female and one-half from the male parent.

After the breaking up of the spireme thread into a definite number of chromosomes, there is formed in the cell the so-called amphiaster. The development and activities of this interesting structure seem to be for the purpose of arranging the chromosomes in position for division. All the processes concerned in the prophases are preparatory to the final division, and ultimate distribution of chromatin to the new cell.

11. Metaphase. — Each chromosome now splits into two exactly equal halves and the two new groups move to opposite sides of the cell. The chromosomes divide lengthwise, and by so doing there results an accurate division of the chromatin into two precisely equivalent portions. Each portion eventually becomes the nucleus of one of the two new daughter cells which result from this division. The most fundamentally important fact about this division of the chromatin is that it is a qualitative as well as a quantitative division. There is much evidence to show that the spireme thread (and, therefore, the chromosomes) is composed of granules or units throughout its length, and each of these units represents a definite character or set of characters in the individual. It follows that when a lengthwise division occurs, these units are divided and a portion of each is passed on to the daughter cells (Fig. 2).

The arrangement of the chromatin in the spireme, its breaking up into chromosomes, and the splitting of the latter into halves are all directed toward the accurate

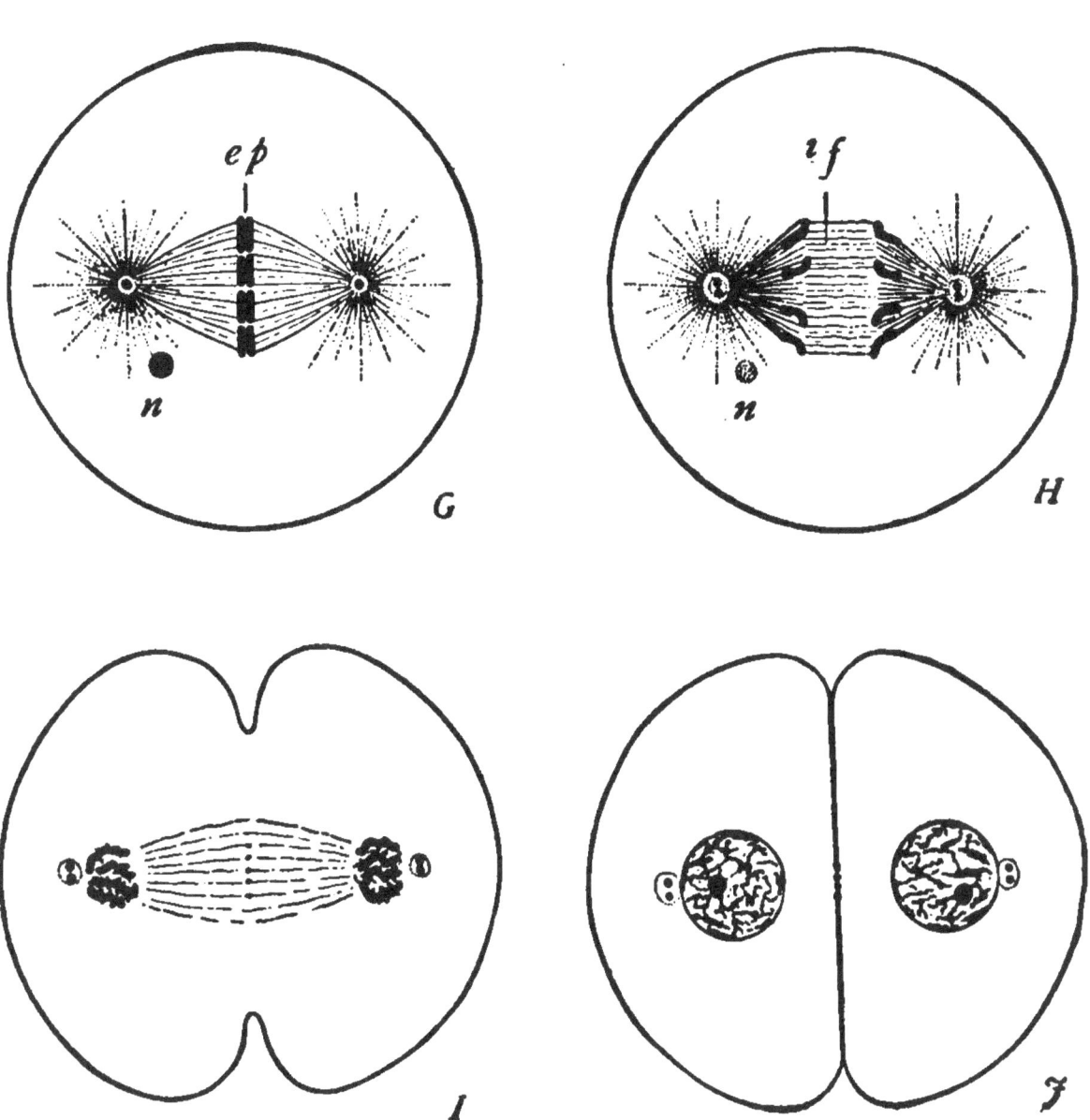

Fig. 2. — Diagram of later phases of cell division. *G*, Splitting of the chromosomes. *H*, Daughter chromosomes diverging to form new cells. *I*, Chromosomes grouped in daughter nuclei, final changes before complete division. *J*, The two new cells.

division of the smaller chromatin granule (chromomere or id as designated by Weismann). In this process, therefore, we have a reasonable physical basis for a better understanding of the transmission of characters from parent cell to daughter cell and thus from parent to offspring.

12. Anaphase. — The essential step in the anaphase is the separation of the two groups of chromosomes which were derived from the splitting of the chromosomes of the parent cell. These pass to opposite sides of the cell, and in many forms these new groups begin at once to assume the appearance of the original nuclear material.

13. Telophase. — Complete division of the cell is finally accomplished by constriction of the parent cell-walls and the formation of a new membrane which encloses the two daughter nuclei each in its own cell. The two groups of chromosomes derived from the parent nucleus are rearranged and become the nuclei of the new daughter cells which in their turn may pass through all the stages described above.

The essential steps, then, in cell division are: (1) The formation of the spireme thread from the chromatin and its division into chromosomes; (2) The splitting of the chromosomes through the middle longitudinally; (3) The movement of the divided portions of the chromosomes to the new or daughter cells.

14. The germ-cells in detail. — In all the higher forms of animals, reproduction is accomplished by the formation of special reproductive cells called the germ-cells. The germ-cells are the product of the reproductive group of cells and are endowed with peculiar powers not generally possessed by the soma- or body-cells. Weismann divided the cells of the body into two very clearly marked and distinct groups, the soma- or body-cells and the germ-cells. The soma-cells are primarily concerned with the individual life of the animal, while the germ-cells are destined solely for the purpose of reproduction. The germ-cells have no important relation to the functional activities which are especially concerned with the individual existence of the animal. They are clearly intended for ultimate separation from the individual and destined to provide for the continuance of the species. In some lower forms of life, the body-cells and germ-cells are not so clearly separated, but in the domestic animals such differentiation is characteristic. The male germ-cell is called the spermatozoön or sperm-cell, and the female germ-cell the egg or ovum.

15. The ovum. — The germ-cell of the female is the egg or ovum (Fig. 3). It is one of the largest cells in the animal body. It is spheroidal in shape and contains a nucleus or germinal vesicle and within the nucleus a nucleolus or germinal spot and a large amount of protoplasm (cytoplasm) surrounding the nucleus. In the protoplasm are often distributed numerous masses of material or yolk, the function of which seems to be chiefly to nourish the fertilized egg-cell. The nucleus of the ovum during the quiescent stage lies near the center of the cell, but gradually moves toward the cell-wall as the egg becomes more mature. During the final stages in the development of the egg, the nucleus loses a large part of its chromatin. This process is called maturation and results ultimately in reducing the number of chromosomes to one-half the number characteristic of the species. This process is preparatory to the fertilization of the egg by the spermatozoön. During fertilization, the chromo-

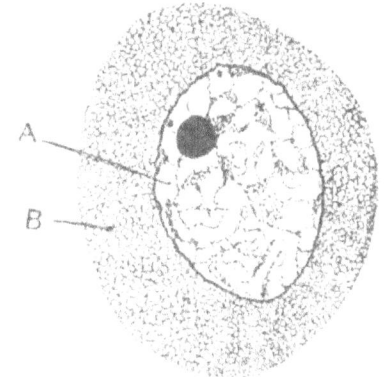

FIG. 3. — The ovarian egg. A, nucleus; B, cytoplasm.

somes of the egg-cell unite with those of the sperm-cell and a new nucleus is thus formed which becomes the active center of the new daughter cell.

16. The spermatozoön.—The male germ-cell or spermatozoön is exceedingly minute and its investigation cor-

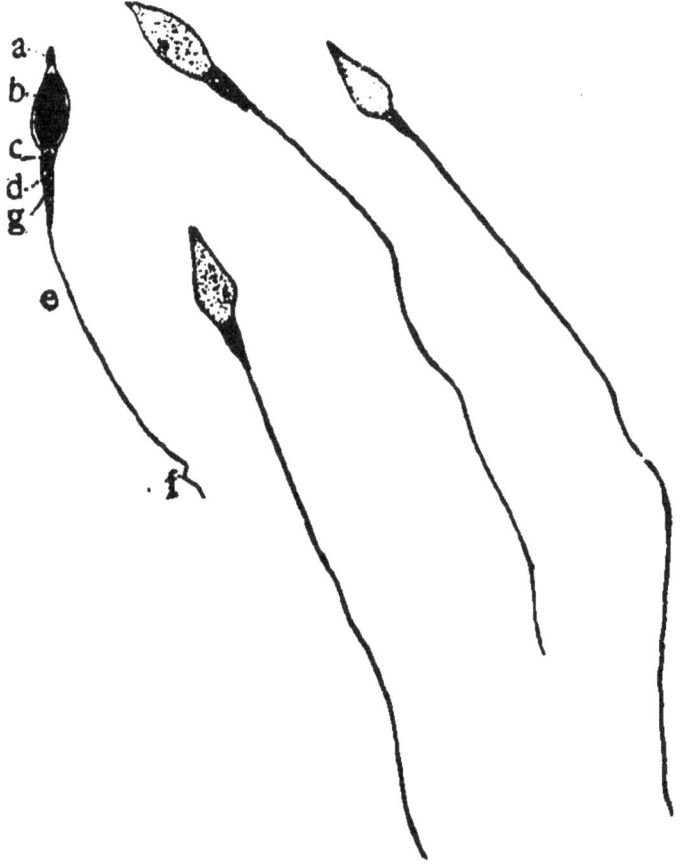

Fig. 4. — Human spermatozoa (greatly magnified). *a*, acrosome; *b*, nucleus; *c*, end knob; *d*, middle piece; *e*, tail; *f*, end piece; *g*, axial filament.

respondingly difficult. Although early discovered, its significance was not clearly determined until 1865 when Schweigger-Seidel and St. George discovered that the sperm was a cell and contained a nucleus, cytoplasm and all the essential elements of a perfect cell. The function of the sperm-cell is to fertilize the egg and thus provide for the reproduction of the species. The spermatozoön in its gross anatomy has the semblance of a very minute tadpole swimming freely about in liquids. It has a distinct head piece and a long slender tail piece or cilium. (See Fig. 4.) When closely examined, the sperm-cell exhibits all the essential characteristics of a typical cell, with a nucleus, cytoplasm and a cell-wall. This cell is so small that in some forms it is but $\frac{1}{100000}$ of the size of the egg-cell. The nucleus occupies almost the entire space available in the head piece, being surrounded by a very thin layer of cytoplasm between it and the cell-wall. The tail of the sperm-cell is joined to the head piece by the middle piece. It is of cytoplasmic origin and possesses the power of motion. This ability to propel itself forward in liquid media seems to be in a way an insurance that the sperm-cell will ultimately approach the egg-cell for the purpose of fertilization.

The power of motion is retained by the spermatozoön for a considerable time under favorable conditions of warmth and moisture. After the sperm enters the egg, it loses the power of motion and the tail piece is absorbed.

The essential portions of the sperm which are concerned in the process of fertilization are the nucleus and the middle piece. Other structures which are accessory to these are the apex by which the sperm attaches itself to the ovum and the tail whose functions have already been described.

CHAPTER II

REPRODUCTION

The value of a breeding animal is measured by its own individual character, its ability to transmit desirable characters to its offspring, and by its prolificacy. Of two animals of equal individuality and inheritance, the one capable of producing numerous and vigorous offspring will be the more valuable. The subject of reproduction in animal-breeding is therefore of great importance and second only to inheritance in estimating animal values.

Two methods of reproduction are common among plants and animals, asexual and sexual. These differ greatly in form and method but accomplish the same ultimate result, which is the continuance of the race.

17. Asexual reproduction. — The asexual method of reproduction is common among the simplest forms of plant and animal life. It is a process of cell division or fission which varies in different forms, but in its most elementary manifestations it is simple cell division. In unicellular forms the cytoplasm of the cell and the nucleus divide into two equal parts and each half becomes a perfect new individual. Each new cell increases in bulk by the absorption of food and in turn becomes a parent cell and reproduces by division as before. A similar form of reproduction is found in the developing embryo of mammals and of the other body-cells. The tissues of the body thus formed cannot, however, continue to divide indefinitely, but sooner or later reach their full development and do not thereafter increase in bulk by further division.

18. Sexual reproduction. — Reproduction in the higher forms of organic life is a complicated process and is accomplished by the activities of a special group of cells which is called the reproductive or generative system. These germ-cells are highly specialized, and their development and functions divide the individuals of a species into two sexes, male and female. The two sexes differ widely in form and characters, but particularly in respect to their physiological relations to reproduction. The product of the male germ-cells is the spermatozoön, an exceedingly small cell but carrying in its minute substance the inherited potentialities of its ancestors. The female germ-cell or ovum similarly derived from the female is much larger, but also carries the germ substance which later, after fertilization by the spermatozoön and under proper conditions, will become a new individual.

19. The reproductive process. — The essential fact in the reproductive processes of all vertebrate animals is the formation of an egg or ovum by the female and of a fecundating fluid containing the spermatozoön or sperm-cell by the male. The union of the two germ-cells is the first step in the independent existence of the new individual. The conjugation of the egg and spermatozoön sets in motion a series of events which, whether viewed from the standpoint of racial significance or of biological interest, has no parallel in the whole realm of biology. Upon the successful development of the fertilized egg-cell depends the future of the species. The union of the egg and spermatozoön takes place in some forms outside

c

the body of the female. In fishes, the spawn containing the eggs is deposited in the water by the female, and the fertilizing fluid of the male containing the spermatozoa is also deposited in the water on or near the female spawn. In all higher animals, including the domestic animals, the fecundating fluid of the male reaches the egg inside the body of the female, and the process of fertilizing the egg is accomplished within the generative organs of the female.

20. Oviparous animals. — In some species of animals, including birds, fishes, most reptiles, and nearly all invertebrates, the eggs are deposited outside the body either before or after fertilization. In the bird family, which includes all the domestic fowls, the egg is fertilized inside the body of the mother, where it undergoes some slight development before being eventually deposited or laid outside the body of the female. Animals of this class are called oviparous, to distinguish them from viviparous forms which bring forth their young alive. The future development of the egg deposited by oviparous animals is dependent upon its being supplied with favorable conditions of heat and moisture. This period of development outside the body is called the period of incubation and is, in many ways, similar to the period of gestation in viviparous animals. Animals in which the fertilized egg develops inside the body are called viviparous, and this development proceeds to a point where the young animal is, in the main, able to carry on a separate and independent existence. The period of growth inside the body of the mother in mammalian animals is called the period of gestation, and during this period the developing embryo is gradually fitted for an independent life outside the protecting body of the mother.

21. Primary and secondary sexual characters. — The particular characteristics which differentiate the male from the female sex may be divided into primary and secondary characters. The primary characters are those which are peculiar to the sex. In the male, the form and functions of the generative organs clearly differentiate him from the female. As Lee [1] has pointed out, the function of the primary sexual characters of the male is the production of spermatozoa and the impregnation of the female egg. The activities of the primary sexual organs of the female center about the production of the egg and the development of the embryo.

The secondary sexual characters are those which may often be possessed in common by both sexes but experience a special and somewhat different development in the two sexes. Examples of secondary sexual characters are the horns of the ram in some breeds of sheep, the spurs of the cock, the tusks of the boar, the various colored plumage of many male birds, and the crest in stallions and bulls. Under certain conditions, the secondary sexual characters which are in general characteristic of the male sex alone may be developed in the female. Thus, hens sometimes develop spurs and ewes develop horns.

Castration of the male causes a check in the development of the secondary sexual characters and causes the male to approach more nearly to the form and appearance of the female.

22. The reproductive organs of the male (Fig. 5).—The organs of generation in the male are: (a) the testicles, corresponding to the ovaries in the female; (b) vasa deferentia, or the ducts leading from the testicles to the ejaculatory

[1] "American Text Book of Physiology," 1903, p. 443.

duct; (c) vesiculæ seminales; (d) prostate glands; (e) Cowper's gland; (f) urethra, through which the urinary and genital secretions are conveyed; (g) the penis, through which the semen of the male is conveyed to the female genital organs. The secretions of the vesiculæ seminales, the prostate, and the Cowper's gland all empty into the urethra, where they mix with the seminal fluid from the testicles. The testicles, vasa deferentia and urethra and penis are often called the essential organs of generation, while the remaining three are referred to as the accessory organs.

23. The testicles. — The essential sexual elements, the spermatozoa, are formed in the testicles. The removal of the testicles, therefore, destroys the ability of a male animal to elaborate the male germ-cells and permanently destroys his fecundity. The structure of the testis is described in some detail by Marshall as follows:[1] "This organ is enclosed within a fibrous capsule, the tunica albuginea, which is very rich in lymphatics. It is covered by a layer of serous epithelium reflected from the tunica vaginalis. Posteriorly the capsule is prolonged into the interior of the testis in the form of a mass of fibrous tissue (the mediastinum testis). Certain other fibrous processes or trabeculæ also project inwards from the capsule, and divide the glandular substance into lobules. The efferent ducts of the testis (vasa efferentia) open into a single convoluted tube situated at the posterior margin of the organ and attached to the mediastinum. This is the epididymis. Its lower extremity is prolonged into a thick-walled muscular tube (the vas deferens) which is the passage of exit for the seminal fluid or sperm-containing secretion. The glandular sub-

[1] Marshall, "The Physiology of Reproduction."

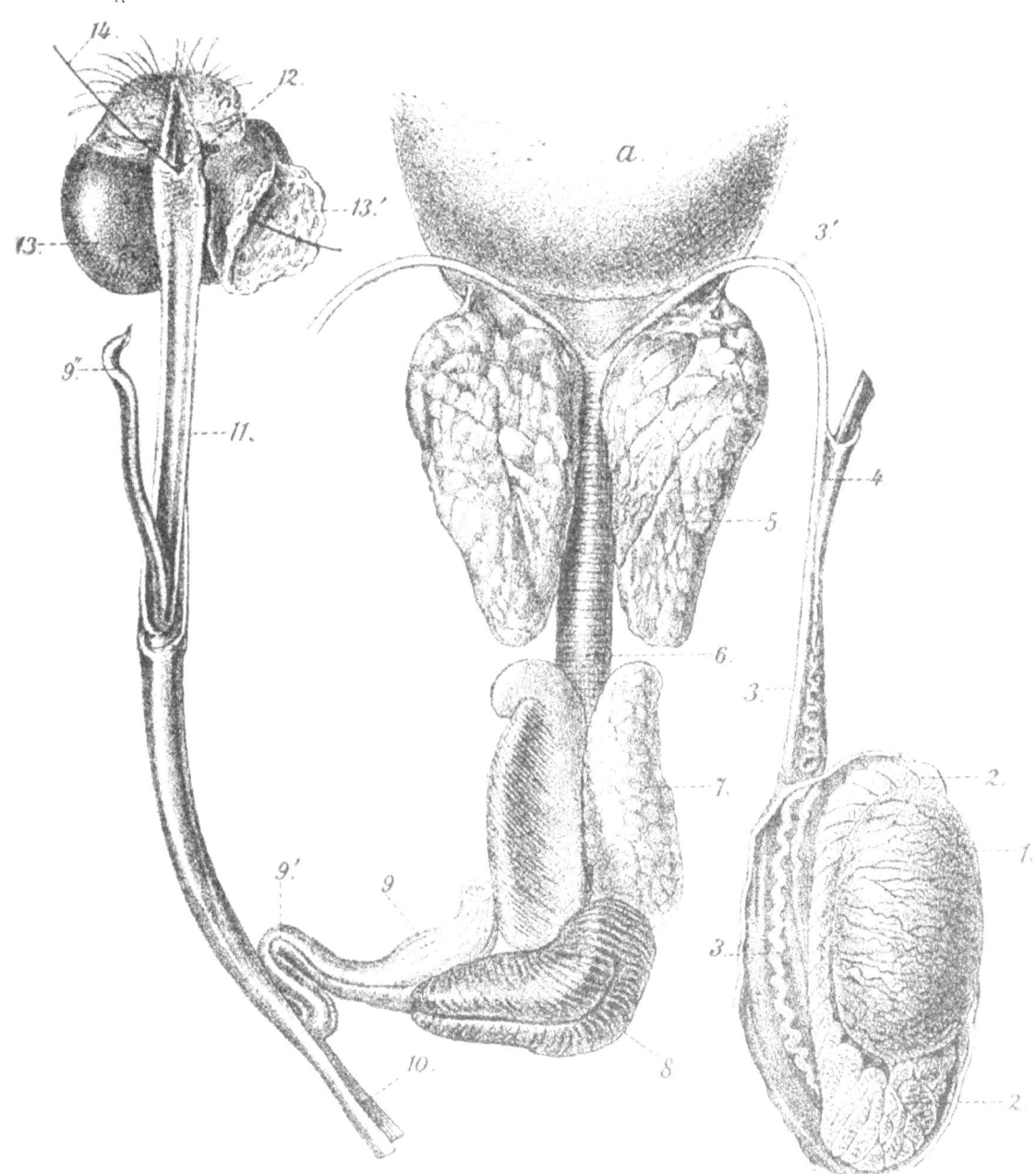

Fig. 5. — Genital organs of boar. 1, testicle; 2, epididymis; 3, vas deferens; 4, spermatic artery; 5, vesicula seminalis; 6, prostate; 7, Cowper's gland; 8, bulbo cavernosus muscle; 9, 9' and 9'', penis; 10, retractor penis muscle; 12, orifice of preputial pouch.

stance of the testis is composed of the convoluted seminiferous tubules, two or three of which join together to form a straight tubule which passes into the body of the mediastinum. The straight tubules within the mediastinum unite in their turn, giving rise to a network of vessels called the rete testis. From the rete the vasa efferentia are given off. Between the tubules is a loose connective tissue containing a number of yellow epithelioid interstitial cells. The connective tissue also contains numerous lymphatics and blood-vessels (branches of the spermatic artery). The nerves of the testis are derived from the sympathetic system, but a few filaments come from the hypogastric plexus."

The male reproductive organs [1] other than the testicles are chiefly concerned in providing for the transmission of the fully mature sperm-cells from the testicles. The seminal fluid is secreted by the seminiferous tubules and after expulsion is mixed with other fluids from the male accessory organs.

The function of the seminal fluid seems to be to provide a favorable medium for the spermatozoa.

24. Castration. — The removal of the testicles of the male results in profoundly influencing not only the breeding function of the animal but in marked bodily change. While the testicles seem to have no important connection with or relation to the body-cells and their removal interferes in no way with the normal, healthy functioning of the other bodily organs, it is nevertheless true that castration materially influences the animal economy.

---

[1] The student is referred to standard works on anatomy and physiology for a detailed discussion of the organization and functional relations of the various male reproductive organs. *See* Sisson, "Veterinary Anatomy."

Chauveau, "Veterinary Anatomy."

If the young male is castrated, the external appearance of the animal undergoes a gradual change. The whole aspect becomes less masculine. The shoulders, neck and crest develop relatively to a much less extent. The hind quarters are relatively better developed in the castrated animal. The temperament and disposition undergo radical changes. The bull castrated before puberty fails to develop the heavy head and horns, curly hair and protruding eye characteristic of this animal. In swine, castration prevents the development of the shoulder plates and tusks. The horns of sheep are very greatly dwarfed, though the same effect is not so marked in cattle. The removal of the testicles also influences the physiological constitution of the animal. It is well known that oxen grow larger and heavier than the uncastrated bulls. The same result is observed in capons which often grow to a much greater weight than the normal male. The castration of old boars results in the disappearance of the strong odor characteristic of the flesh, and their food value is thereby increased.

The removal of the ovaries (spaying) of the female is followed by phenomena similar to those observed in the castrated male. The spayed female loses her feminine appearance and approaches the male in general character. Like the male, the disposition becomes quieter and the general physiological condition of the animal favors more rapid laying on of fat.

At the Massachusetts Experiment Station, Goodale[1] castrated a brown Leghorn cockerel at twenty-four days old. At the same time the fresh ovaries of two brood sisters were cut in several pieces and placed under the skin of the castrated cockerel. The bird developed in

---

[1] Goodale, "Science," Vol. 40 (1914), p. 549.

every external character like a female, so much so that expert poultry men were deceived.

We may conclude that the castration of animals is successful in preventing undesirable animals from reproducing, improving the fattening qualities of the meat animals, increasing the size in some cases, improving the quality of the flesh, and increasing the value of draft animals.[1]

25. The reproductive organs of the female (Plate II). — The essential organs of reproduction in the female are the two ovaries. The accessory organs are the Fallopian tubes, the uterus, the vagina, the vulva and the mammary glands. The function of the female organs of generation is the production of the female germ-cells or ova and, when fertilized, to provide nutrition and proper support for the developing embryo.

26. The ovaries. — The female egg has its origin in the ovary. This organ is composed of connective tissue, blood-vessels, nerves and lymphatics enclosed in an outer covering — the epithelium. A cross-section of the ovary of any mature breeding animal will exhibit a large number of follicles or sacs scattered through its vascular substance (Fig. 6). The latter are small, varying from one-hundredth to one-thirtieth of an inch in size.[2] These small sacs are called the Graafian follicles, and each contains an ovum or egg. The eggs or ova in these follicles exhibit various stages of development, some being almost or quite mature, while others are very small and undeveloped. The beginning of the formation of a Graafian follicle is indicated by a slight depression in the surface of the ovary which gradually extends into the substance of the ovarian

[1] Pusch, "Allegemeine Tierzucht," 1911, p. 153.
[2] Smith, "Physiology of the Domestic Animals," p. 909.

tissue in the form of a tubule. Eventually this sac becomes constricted at the surface of the ovary until finally the external opening is entirely closed and the tubule becomes a closed follicle within the tissues of the organ. Within the follicle itself, there have been formed, in the meantime, single, large spherical cells (primordial germ-cells) from which one or sometimes two ova are developed. The part of the Graafian follicle not occupied by the ovum is filled with a fluid substance.

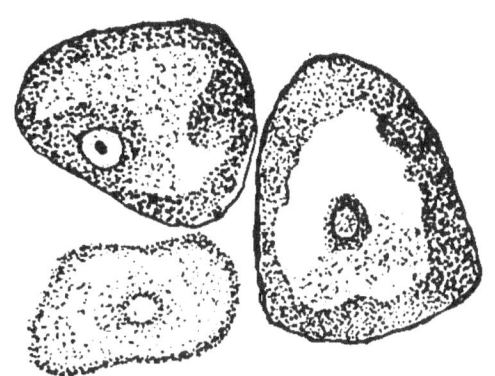

Fig. 6. — Sections through ovary of rat, typical of ovarian structure in mammals.

The follicle increases in size and approaches the center of the ovary until near maturity, when it rises to the surface and finally is ruptured, thus liberating the enclosed egg.

The bursting of a Graafian follicle and the discharge of the egg is called ovulation. This event is marked by certain phenomena indicating increased sexual activity. It is believed that menstruation or the period of heat in domestic animals is coextensive with the ripening of the egg. It is true, however, that, under some circumstances, ovulation may occur before or after the period of heat. The ripening of the first Graafian follicle in general marks the beginning of puberty, but has been known to occur even in infancy. In animals generally ovulation does not occur during pregnancy, but there are numerous exceptions to this rule, as will be described later in the case of some animals which have come in heat and have even conceived again, although pregnant at the time (Fig. 7).

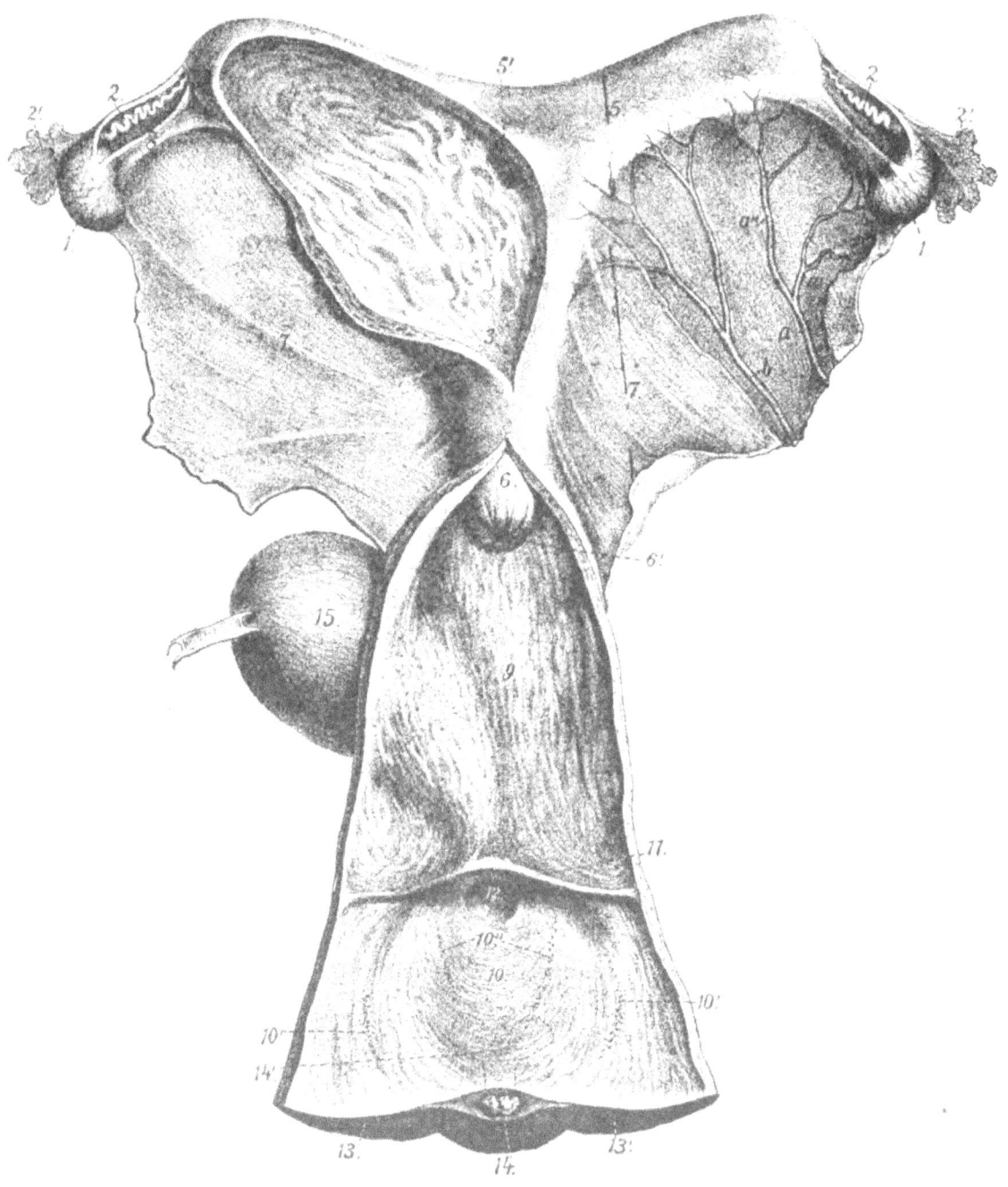

Fig. 7. — Genital organs of mare. 1, ovary; 2, Fallopian tube; 2', fimbriated end of Fallopian tube; 3, uterus; 5, horn of uterus; 6 and 6', cervix; 7, broad ligament of uterus; 9, vagina; 10, vulva; 13 and 13', lips of vulva; 14, clitoris; 15, urinary bladder; *a*, utero-ovarian artery with branches to ovary (*a*) and uterus (*a''*); *b*, uterine artery.

Some of the domestic animals do not come in heat while suckling young, while others discharge eggs and undergo the periodical phenomenon of heat as readily during lactation as at any other time.

27. The Fallopian tubes. — The egg which has been discharged as the result of the processes described above finds its way to the uterus through the Fallopian tubes in mammals or the oviduct in birds. These are small, very crooked canals leading from the ovary to the uterus. This accessory organ is not rigidly joined to the ovary by tissues, but the end nearest the ovary is mostly free to move to different sides of the ovary. The ovarian end of the Fallopian tube is expanded into a fimbriated extension which spreads out not unlike the fingers of the hand. The comparison will be still more exact if we conceive of the fingers of the hand as being connected by web-like tissues.

In many mammals, the tube is lined with cilia which move from the ovary toward the uterus. In normal cases when the egg is discharged from the ovary, the fimbriated expansion of the Fallopian tube clasps the ovary at the point where the Graafian follicle bursts through its walls. The ciliary movement within the tube, assisted by muscular movements of the tube itself, carries the egg from the ovary to the uterus. The time required for the passage of the egg through the Fallopian tube has not been definitely determined for all mammals, but is known to vary from three to eight days.

The union of the spermatozoön and the egg usually takes place in the Fallopian tube. To accomplish this union, it is necessary for the sperm-cell to pass into the uterus and up into the tube. This it is able to do by reason of its possessing the power of independent motion.

28. The uterus. — The uterus is a muscular sac connecting the Fallopian tubes and the vagina in which the development of the fertilized egg is carried forward until expelled from the body of the mother at the time of parturition (Figs. 8, 9, 10).

In many mammals the uterus divides into two tubes called horns. Each horn is connected with the corresponding ovary by means of the Fallopian tube.

The portion of the uterus nearest the vagina is somewhat constricted to form the cervix or neck of the uterus. The vagina connects the uterus with the external genitals called the vulva.

29. The mammary glands. — The possession of mammary glands whose function is the elaboration of food materials for the young offspring is characteristic of all mammals. These glands are highly developed and functional in the fertile female, but are also present in rudimentary form in the male. In rare cases the rudimentary glands present in the male have been known to function. Hayward at the Delaware Experiment Station reports the case of a registered Guernsey bull owned by that institution whose mammary glands were developed to the extent of producing a small amount of milk. Milk has also been produced from the rudimentary mammary glands of male goats and sheep. In man the rudimentaries of males have produced milk at birth and at puberty and in exceptional cases at other times. The number of nipples in a species bears some relation to the normal number of young produced in a litter, and also to the needs of the young animal. The glands are generally arranged in pairs either along the ventral side of the thorax or abdomen.

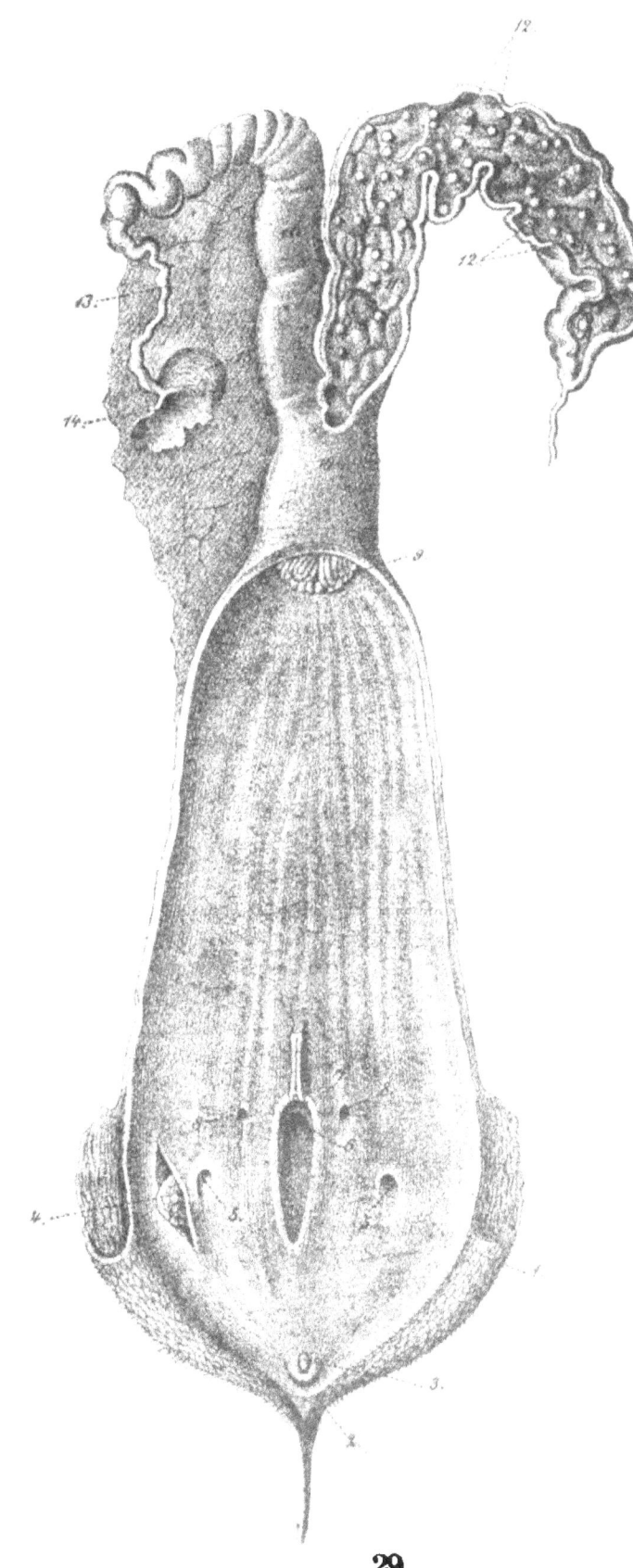

FIG. 8. — Genital organs of cow. The right horn of uterus, the vagina and vulva opened to show interior. 1, lips of vulva; 3, clitoris; 4, glands of Bartholin; 6, suburethral diverticulum (blind pouch); 7, external openings of urethra; 9, neck of womb; 10, body of uterus; 11, horn of uterus; 12, cotyledon; 13, Fallopian tube; 15, ovary.

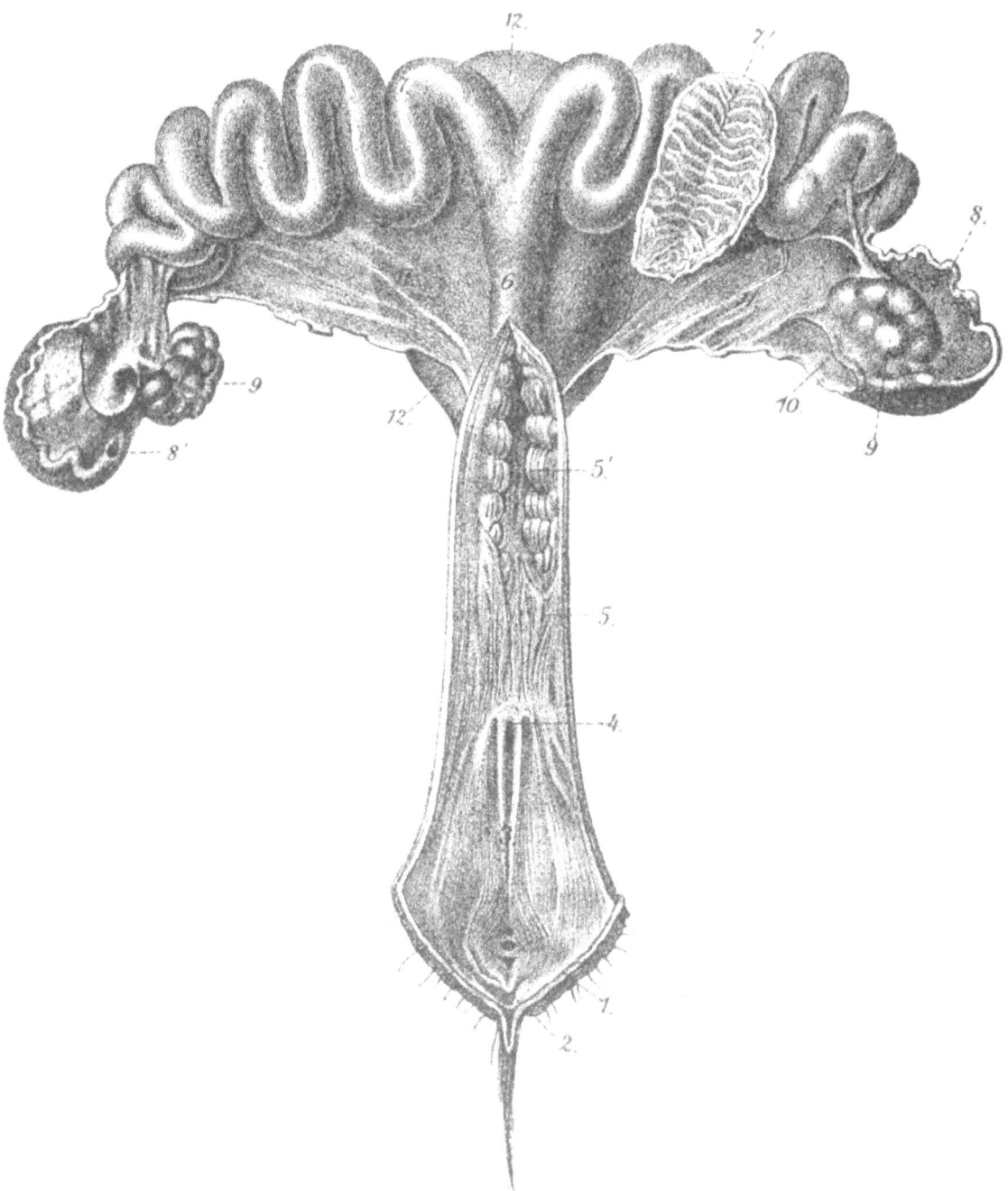

Fig. 9. — Genital organs of the sow. 1, lips of vulva; 2, clitoris; 3, vulva; 4, external opening of urethra; 5, vagina; 5, cervix; 6, body of uterus; 7, horn of uterus; 7', horn of uterus opened to show interior; 8, Fallopian tube; 8', abdominal opening of Fallopian tube; 9, ovaries; 12, urinary bladder.

REPRODUCTION

30. Structure of mammary glands.[1] — The mammary glands are made up of lobes which in turn are further divided into lobules. The latter arise from secretory alveoli. The lobule is chiefly connective tissue binding together the milk ducts. The alveoli unite together to form the lactiferous ducts which open externally. These ducts are provided with reservoirs wherein the milk is accumulated during the period of active lactation. During lactation the alveoli secrete milk. This secretion goes forward at all times, but is particularly active during suckling. The milk drawn first is of poorer composition in respect to solids than that drawn near the end of the milking. This may be due to the fact that the larger globules pass through the ducts with greater difficulty and are thus retained longer in the gland.

The processes concerned in milk secretion are not entirely understood, at least three views having been held. One hypothesis is that the secretory cells themselves break down and thus set free their contents as is the case with the sebaceous glands. Another view is that the milk is simply excreted from the cell without causing the breaking down of the cell, in a manner similar to that which occurs in many secretory glands.

"The third theory[2] was first suggested by Langer, and has since been adopted, with various slight modifications, by Heidenhain, Steinhaus, Brouha and others. According to their view the cells of the gland lengthen out, so that their ends come to project freely into the lumina of the alveoli. The projecting portions then undergo a process of disintegration before or after

[1] Marshall, "The Physiology of Reproduction," p. 553.
[2] *Ibid.*, p. 560.

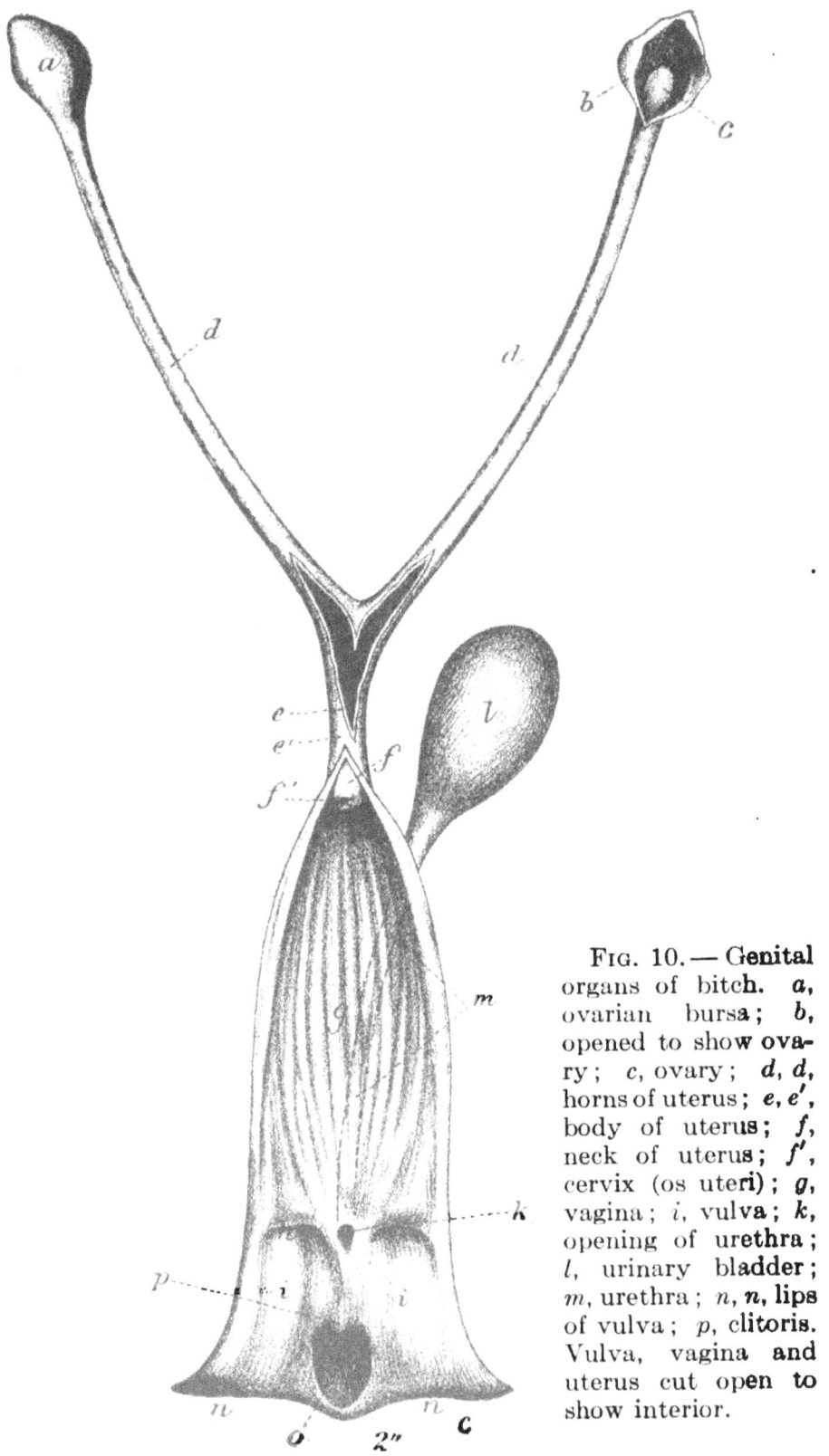

Fig. 10. — Genital organs of bitch. *a*, ovarian bursa; *b*, opened to show ovary; *c*, ovary; *d, d*, horns of uterus; *e, e'*, body of uterus; *f*, neck of uterus; *f'*, cervix (os uteri); *g*, vagina; *i*, vulva; *k*, opening of urethra; *l*, urinary bladder; *m*, urethra; *n, n*, lips of vulva; *p*, clitoris. Vulva, vagina and uterus cut open to show interior.

becoming detached, and the cell substance passes into solution to form the albuminous and carbohydrate constituents of the milk. The fat droplets which collect in the disintegrating part of the cell give rise to the milk fat. The basal portions of the cell remain in position without being detached, and subsequently develop fresh processes, which in their turn become disintegrated. It is believed, however, that some cells simply discharge their fat droplets and other contents into the lumina, while otherwise remaining intact."

31. Fertilization of the ovum. — We have seen how growth may continue for a long period by successive cell division. Indeed, in many of the simpler forms. Wilson has pointed out that " as far as we can see from an *a priori* point of view, there is no reason why, barring accident, cell division should not follow cell division in endless succession in the stream of life."[1]

In some of the very simplest forms, no sexual union has so far been discovered. But, under normal conditions, reproduction without sexual union is rare and practically unknown in the higher animals. The impetus to growth and cell division is not permanent and must be constantly renewed. The stimulus to renewed growth is accomplished by the definite mixture of living protoplasm from two entirely distinct individuals. This process of fertilization involves a union of the ovum of the female and the spermatozoön of the male. This union is the beginning of the life of each individual and results not only in energizing the protoplasm of the germ-cell, causing it to divide and grow, but the admixture of germ material from two different individuals introduces into the new organism two distinct lines of inheritance.

[1] Wilson, "The Cell."

32. The nature of fertilization. — The essential nature of fertilization still remains a matter of discussion. It may be that the fertilization of the egg is primarily a rejuvenescence of the protoplasmic material which, for some at present unknown reason, has lost the power of further growth through cell division. This view was held by Butschli, Hertwig, Minot, Engelman and others. Even in the simpler forms of life where anything like sexual union is absent, the cycle of growth is continually reinaugurated by the conjugation of independent cells. In all higher forms the egg is stimulated to growth and cell division by the introduction of the sperm.

Weismann has looked upon fertilization as a source of variation and has maintained that this should be regarded as the chief function of this process. Both the theory of rejuvenescence and that which regards this process as chiefly a source of variation are in accordance with the observations of practical breeders. Cross-breeding is known to induce greater vigor and increased fecundity and at the same time to break up the fixed characters of the breed or type. It is equally well established that under certain conditions in-breeding tends to identity of character and results in sterility and weakness of constitution. But, after all, the real purpose and nature of sexual union in reproduction is still an unsolved problem.

33. The process of fertilization. — The ultimate purpose of the sexual union of animals is to insure the fertilization of the egg by the spermatozoön. When the egg and sperm meet within the generative organs of the female, significant and important changes are set in motion, which eventually result in an admixture or union of the germ substance of the two parents. These changes are primarily concerned with the nuclei of the egg and the sperm.

In most species of animals, there is a definite attraction existing between the egg and the sperm-cell, which causes the spermatozoön to attach itself to and finally penetrate the egg. This attraction is probably of a chemical nature. Pfeffer found that solutions of malic acid were as successful in attracting the sperm-cells of ferns as the substance which was thrown off by the female sex cells. In other experiments, other chemical substances have exhibited a specific attraction for the sperm. The seat of this chemical substance which pulls the sperm to the egg seems to be located in the cytoplasm of the egg and not in the nucleus.

The point at which the spermatozoön enters the egg is often predetermined by the existence of a depression or opening (micropyle) in the wall of the ovum. In some cases, there exists a special protoplasmic attraction cone at which point the sperm enters the egg. The entrance of the sperm in most cases is followed rapidly by the formation of a vitelline membrane which surrounds the egg and prevents the entrance of other spermatozoa.

Normally in mammals one sperm only enters the egg. When through accident two or more spermatozoa enter the substance of the egg, the developmental changes are abnormal and the daughter nucleus soon dies.

34. The chromosomes. — When the germ nuclei unite to form the daughter nucleus of the new cell, it seems probable now that the chromatin substance does not fuse in true fashion but the chromosomes may lie side by side within the nucleus of the new cell. It is not definitely determined that these chromosomes remain thus separate and apart throughout the life of the cell, but such may be the case. After the reduction of the chromosomes in the maturation of the germ-cells and accompanying the formation of the polar bodies, the number of chromosomes is reduced to one-half the number existing in the body-cells and these uniting with an equal number from the sperm constitute the normal number characteristic of the tissue cells. As Wilson says, "We have thus what must be reckoned as more than a possibility that every cell in the body of the child may receive from each parent not only half of its chromatin substance, but one-half of its chromosomes, as distinct and individual descendants of those parents."

35. Results of union of egg and sperm. — Extraordinary changes follow immediately the physical union of the egg and spermatozoön. These changes are of the most fundamental significance in connection with discussions of development and inheritance.

36. Changes in the ovum. — The entrance of the sperm seems to exert an influence which permeates the entire constitution of the egg substance. Various and important changes now take place, ending finally in an exact union of the germ substance of the two sex cells, thus forming the new or daughter cell. The daughter cell is the beginning of a new individual and becomes the offspring of the parents, from which the sperm and ovum were derived. When the sperm enters the egg, the vitelline membrane is thrown around the outside of the ovum. The nucleus of the egg, which is now called the germinal vesicle, moves to the wall, and changes occur which result in the formation of the polar bodies. This process prepares the germ nucleus of the egg for fertilization.

The polar bodies are formed by successive divisions

of the egg nucleus. During their formation, the number of chromosomes characteristic of the species is reduced by half, so that when the egg nucleus is finally ready for union with the sperm nucleus, it contains exactly one-half the number of chromosomes usually present in the cell. There are formed in all three polar bodies. The divisions which occur in their formation separate the mass of chromatin originally present in the germinal vesicle into four equal parts. One-fourth part enters the egg nucleus and the other three parts are distributed, one to each of the polar bodies. The polar bodies are in no direct way concerned in fertilization and soon disintegrate and disappear.

37. Changes in the spermatozoön. — After contact with the egg, the tail of the sperm soon degenerates either outside or, in some cases, inside the egg.

The nucleus of the sperm grows rapidly in size. The nucleus is further stimulated to cell division by the influence of the cytoplasm of the egg.

38. The significance of reduction. — All the phenomena attending the processes of fertilization seem to have been specifically arranged for the purpose of bringing about a reduction of the number of chromosomes in the germ-cells to one-half that found in the other cells of the body. A result so universal in plants and animals must possess some significance in the reproduction of living forms. This change is characteristic of the germ-cells only and always occurs prior to and in preparation for the union of the spermatozoön and the egg. What is the real significance of the reduction of the chromosomes? Is this reduction of the chromatin a quantitative one, or is it in some way a qualitative division? It must be admitted that we cannot yet give positive answers to these questions. That this process is one of fundamental significance and is intimately connected with the problem of how characters are transmitted, is certain. Manifestly its purpose is to provide for a constant number of chromosomes in the body tissues. It is not, as some have maintained, a mere mass reduction of the chromatin. Weismann's ingenious theory of the germ-plasm attempts to explain the process and to point out the hereditary significance of reduction. In an earlier investigation, Roux held that the hereditary qualities are represented by the individual chromatin granules. During cell division, these arranged themselves side by side in the spireme thread, and the splitting of the spireme thread longitudinally actually resulted in halving each individual chromatin granule. Quoting Wilson,[1] — " Roux assumes, as a fundamental postulate, that division of the granules may be either quantitative or qualitative. In the first mode, the group of qualities represented in the mother granule is first doubled and then split into equivalent daughter groups, the daughter cells, therefore, receiving the same qualities and remaining of the same nature. In 'qualitative division,' on the other hand, the mother group of qualities is split into dissimilar groups, which, passing into the respective daughter nuclei, lead to a corresponding differentiation in the daughter cells. By qualitative divisions, occurring in a fixed and predetermined order, the idioplasm is thus split up during ontogeny into its constituent qualities which are, as it were, sifted apart and distributed to the various nuclei of the embryo. Every cell nucleus, therefore, receives a specific form of chromatin which determines the nature of the cell at a given period in its later history. Every cell is thus

---

[1] *Loc. cit.*

endowed with a power of self-determination which lies in the specific structure of its nucleus, and its course of development is only in a minor degree capable of modification through the relation of the cell to its fellows."

Weismann conceives that the chromatin (idioplasm) of the germ-cell exists in the form of minute particles which combine to form aggregates, and these again unite to form compound groups and so on until finally we have the chromosomes. He calls the smallest groups determinants, the next larger groups ids (chromatin granules), these in turn combining and forming idants or chromosomes. Weismann in explanation of his theory says: "Ontogeny depends on a gradual process of disintegration of the id of germ-plasm, which splits into smaller and smaller groups of determinants in the development of each individual. Finally . . . only one kind of determinant remains in each cell, viz. that which has to control that particular cell or group of cells. In this cell it breaks up into its constituent biophores and gives the cell its inherited specific character."

In developing this theory it is necessary to assume a very stable condition of the germ-plasm. For example, one must assume that some portion of the original germ-plasm is passed on to the germ nucleus unchanged in structure from generation to generation.

The theories of both Roux and Weismann are in part only based upon demonstrated phenomena occurring in the cell. Some of the most fundamental and far-reaching postulates of these theories are highly speculative and cannot be demonstrated by any known method of research. It must be admitted, however, that Weismann's theory of the germ-plasm with some modifications comes nearer to an adequate explanation of the changes which can actually be observed in the cell during the fertilization and maturation of the germ-cells than any other so far advanced.

39. The origin of the germ-cells. — The origin of the germ-cells and the phenomena attending the process of the reduction of the chromosomes are of great fundamental significance. The functions of the chromatin in modern theories of heredity, the particular meaning of the processes which result in reducing the number of chromosomes to one-half that found in the body-cells, are problems of the greatest interest in modern biology.

The maturation of the germ-cells is brought about by similar processes in the egg and sperm. The important result which is the reduction of the number of chromosomes is accomplished in each. This is brought about finally in the last two maturation divisions resulting in four cells, each of which contains but one-half the number of chromosomes characteristic of the same cells. In the female three of the four cells are the polar bodies which are abortive and disappear. The remaining cell is the ovum and becomes the carrier of the hereditary substance of the female. In the male the reduction divisions occur as in the female, but all four cells are functional and may take part in the process of fertilization.

40. Maturation and reduction in the female (oögenesis). — The female germ-cells are derived from the primordial germ-cells of the mother. Successive divisions of the primordial germ-cells result in the development of a number of cells known as oögonia.

From these the ovarian eggs are directly derived. Their growth is characterized by an increase in the size of the nucleus which in the ovarian egg (oöcyte) becomes the germinal vesicle. Food materials develop in the cytoplasm, which is now called the yolk. During all the changes described above, the number of chromosomes in the ova remains the same as in the body-cells. When the egg undergoes the final preparation for fertilization, as we have seen, the number of chromosomes is reduced by one-half.

41. Reduction in the male (spermatogenesis). — The changes which bring about the reduction of the chromatin in the male germ-cell are almost exactly similar to those which have been described in the development of the ovarian egg. The spermatozoa originate in the primordial germ-cells. In their earlier development by cell division, numerous spermatogonia are formed which for a time continue to divide with the typical number of chromosomes found in the soma-cells. In time the spermatogonia cease to divide further and become larger. At this stage they are physiologically equivalent in function to the oöcyte of the female and are known as spermatocytes. There now occur two divisions resulting in four cells, each having but one-half the typical number of chromosomes. Unlike the reducing process in the female which results in only one perfect ovum and three abortive cells, all four sperm-cells are functionally perfect.

42. The period of the œstrum or heat. — In all the domestic mammals, the ripening of the first egg is associated with the first appearance of heat. It is the first evidence of puberty and is accompanied by the rapid development of all the secondary sexual characters distinctive of sex. It is by no means certain that the stimulus which causes the heat or œstrum in animals originates in the ovary. It is possible that the stimulus to egg formation is to be found in the influences set in motion by the œstrum itself. The production of ova and the heat period are so closely associated that the stimulus, whatever it may be that causes the one, will probably under normal conditions directly or indirectly cause the other.

Heape has maintained that since it is known that, in various animals, either menstruation or œstrus may take place without ovulation, and that ovulation may occur without the coincidence of menstruation (Leopold and Mironoff, 1894) or of œstrus (fat), the possibility of isolating these functions is demonstrated. Thus it is no longer impossible to suppose that, while they are both due to similar stimulating influences, one of them may be developed in excess of the other.[1]

It is probable that heat may sometimes occur without the production of an egg, and it is possible that the production of an egg may not always be accompanied by heat, but when such a condition exists, it is to be regarded as the exception and not the rule. It is very clear that the œstrum and ovulation are influenced by nutrition. An insufficient supply of food, deficient in the essential elements required for the normal development of animals, retards the first appearance of heat in young animals and causes irregular periods in mature animals. Breeders of live-stock have long known that the œstrum can be materially influenced by the method of feeding. Skillful stockmen feed the females in such a manner as to cause them to be " gaining " at mating time. This is accomplished by richer feeding or turning to fresh grass.

43. Artificial insemination. — The transfer of the semen of the male to the uterus of the female by the aid of instruments or capsules is known as artificial insemination or, more commonly among practical breeders, as

---

[1] Heape, *loc. cit.*, p. 34.

artificial impregnation. The artificial insemination of the domestic animals, particularly cows and mares, has been practiced in various parts of the world for many years. As early as the time of Spallanzani [1] (1784) artificial insemination was successfully accomplished in dogs. It seems probable also that the Arabs have been familiar with the possibilities of this practice for centuries.[2] The insemination of mares, cows and bitches as a remedy for sterility has been demonstrated by Huish.[3] The Russian investigator Iwanoff [4] was successful in inducing pregnancy in rabbits and guinea pigs by artificial means. Artificial insemination is employed for the purpose of overcoming certain forms of sterility in mammals, specifically, constriction of the muscles surrounding the neck of the uterus. It is also successful in cases of acid secretions of the vagina which are unfavorable to the proper functioning of the sperm-cells after they have been deposited in the generative organs of the female. Artificial insemination is also a practical method of extending the usefulness of a valuable male, as by this means one male may be used successfully for breeding a much larger number of females.

44. Methods of artificial insemination. — The most common methods in use are insemination with a specially made syringe or the introduction of the semen in capsules. By one method the semen is collected from the vagina of the mare after the service of the stallion by introducing the syringe into the vagina until the mouth of the syringe

[1] Spallanzani, "Dissertations," vol. II, 1784.
[2] Gautier, "Le Fécondation Artificielle," Paris, 1889.
[3] Huish, "The Cause and Remedy of Sterility in Mares, Cows and Bitches," London, 4th edition, 1899.
[4] Iwanoff, "De la Fécondation Artificielle chez les Mammifères," *Arch. des Sciences Biologiques*, vol. XII, 1907.

is immediately over the cup-shaped depression just in front of the cervix. The semen is then drawn into the syringe and the instrument introduced into the vagina of the mare to be artificially bred and pushed carefully through the neck of the womb to insure the depositing of the semen inside the uterus. Care should be taken not to introduce the point of the syringe into the opening to the bladder which is only five or six inches from the external opening of the vagina. The transfer of the semen to the waiting mare should be made without unnecessary delay and all instruments should be kept at body temperature during the operation. All instruments should be thoroughly sterilized in hot water before using.

Many interesting questions of biological and practical interest are raised by the practice of artificial insemination. How long after the semen is collected will it continue to be potent? What external conditions such as cold, heat, light and air affect the vitality of the germ? How long does the semen retain its vitality within the reproductive organs of the female? Partial answers to these questions have been given through the investigations of Lewis.[1]

45. Conditions influencing the vitality of the sperm-cells. — The vitality of semen collected and preserved at different temperatures under laboratory conditions showed great variations. High temperatures were generally unfavorable. At the end of one hour the percentage of semen which was alive and active was: at 33° C., 40 per cent; at 30° C., 45 per cent; at 26° C., 85 per cent, and at 18° C., 90 per cent. At the end of two-and-one-half hours all the sperm-cells maintained at the temperatures of 33° C. and 39° C. were dead, while 65 per cent of the

[1] Lewis, Oklahoma Experiment Station, Bulletins 93 and 96.

cells were alive at the temperature of 18° C., and 45 per cent of the cells kept at a constant temperature of 26° were still active. The sperm-cells from boars of several breeds showed similar behavior and in every case the higher temperatures were unfavorable. Semen kept at a temperature of 31 to 32° C. and exposed to the diffused light of the laboratory resulted in the death of practically all the sperm-cells at the end of seven hours. A portion of the same semen protected from the light by wrapping in black paper showed 40 per cent of the germ-cells alive at the end of the same time. The spermatozoa exposed to direct sunlight for ninety minutes were all killed, while the portion protected from direct rays of the sun showed 80 to 90 per cent alive at the end of the same time.

46. Effect of too frequent breeding on the sperm-cells. — Lewis[1] found that the number of sperm-cells in the semen collected from the first service of a vigorous stallion was 428,000 to a cubic millimeter. The stallion was permitted one service daily for nine days. The number of sperm-cells diminished rapidly until there were only 74,300 sperm-cells to a cubic millimeter at the ninth service. The continuous and frequent service of the stallion also resulted in weakening the vitality of the sperm-cells. The semen from the first service kept at constant temperature of 13 to 21° C. showed twenty-five per cent of the sperm-cells alive after six-and-one-half hours. The sperm-cells in semen collected from the ninth service showed only five per cent of the cells alive after six hours. Simple exposure to the air seemed to have no deleterious effect on the vitality of the spermatozoa.

The standard of judging of the vitality of the sperm-

[1] *Loc. cit.*, p. 35.

cells in this investigation was microscopic evidence of normal motion and as pointed out by Lewis this method does not necessarily measure the ability of the sperm successfully to fertilize the ovum. The addition of water to semen seems to lower the vitality of the sperm-cells. The presence of urine also has a retarding influence on the activity of the spermatozoa. Sperm-cells kept in contact with rubber lose their vitality more quickly than when preserved in a glass retainer.

47. Vitality of spermatozoa within the female generative organs. — How long do the sperm-cells retain their vitality after being deposited in the generative organs of the female? The answer to this question is of practical importance, as it has an important bearing upon the particular time or stage during the heat at which the union of the male and female will be most likely to result in offspring.

The period during which perfectly healthy sperm-cells retain their vitality and power of motion under laboratory conditions is comparatively short. And while the conditions for a longer period of vitality are presumably much more favorable inside the generative system of the female, yet investigations on mares and sows [1] seem to point to the fact that the life of the sperm-cells in the uterus of the female is comparatively brief. This is contrary to the opinion of many practical breeders. It is generally believed that the spermatozoa retain their vitality in the reproductive organs of the female for a number of days. It is stated by some veterinary authorities [2] that the spermatozoa will live in the vagina or womb of the mare from six to twelve days under the

---

[1] Lewis, Oklahoma Experiment Station, Bulletin 96.
[2] *Breeder's Gazette*, vol. 43, 1903, p. 683.

most favorable conditions. A normal alkaline solution in the uterus is the most favorable medium for the long continued vitality of the sperm-cells. It is very often the case that the secretions in the reproductive organs of the mare are acid and such a chemical condition is very unfavorable to the continuance of the life of the sperm. It is true that the sperm-cells will live but a few hours in such an acid medium as is sometimes found in the uterus. A study of the literature on the longevity of the sperm-cells in the female reproductive organs is somewhat confusing. It is probable that very considerable differences exist in the vitality of the spermatozoa from different individuals, but even this is scarcely sufficient to explain the wide discrepancies reported by various investigators. Various authors [1] have reported the presence of live and motile spermatozoa in the uterus of dogs eight days after coition. Marshall and Jolly [2] found live sperm-cells in the vasa deferentia of the rabbit ten days after the removal of the testes, but all were dead at thirteen days. The sperm-cells of bats are reported by Benecke and others to be deposited in the female organs in the autumn, there to remain dormant until the following spring. Ovulation is induced by the warm weather of early spring and the spermatozoa which have lain dormant throughout the hibernating period become active and insemination occurs. In the domestic hen, according to Lillie,[3] "The period of life of the spermatozoa within the oviduct is considerable as proved by the fact that hens may continue to lay fertile eggs for a period of

[1] Hertwig, "Handbuch der Entwicklungslehre."
[2] Marshall and Jolly, "The Œstrus Cycle in the Dog." *Phil. Transactions.* B., vol. 198, 1905.
[3] Lillie, "The Development of the Chick," 1908, p. 35.

at least three weeks after isolation from the cock. After the end of the third week the vitality of the spermatozoa is apparently reduced, as eggs laid during the fourth and fifth weeks may exhibit at the most abnormal cleavage, which soon ceases. Eggs laid forty days after isolation are certainly unfertilized and do not develop."[1] That the spermatozoa may continue to fertilize eggs in the hen for at least twenty days after coition is noted also by Spallanzani.

The researches of Lewis [2] including records of twenty-five sows showed that in three cases only were the sperm-cells alive and active after twenty hours existence in the female organs of generation. In two cases live spermatozoa were collected from the uterus at the end of forty hours. In most cases the sperm-cells taken from the female organs were all dead at the end of sixteen hours after the sow had been bred to a healthy normal boar. The rupture of the Graafian follicles was found to occur almost universally during the last part of the heat. "In no case were the follicles found ruptured during the first twenty-four hours of heat and in most of the cases a period of thirty hours elapsed after the first signs of heat before many of the egg-cells escaped from the ovary." In no case were the Graafian follicles found ruptured in sows which were examined early in the heat. In one sow ovulation did not occur until forty-five hours and in another case seventy hours after the beginning of heat. From the fact that in swine the duration of the vitality of the sperm-cells in the generative organs is so short and that ovulation occurs during the last part of the heat, it is apparent that sows should be bred during the last part of the heat. To be more explicit, the sow should

[1] Spallanzani, "Dissertations," vol. II, 1784.
[2] *Loc. cit.*, p. 7 *et seq.*

be bred not less than thirty hours after the beginning of heat. In practice, the animal-breeder may safely assume that the normal active existence of the sperm-cells in the uterus of mammalian animals is short and that therefore to insure successful conception, the actual service of the male should occur very near to the time when the heat is at its height.

48. Effect of intoxication of the male parent on his offspring. — The fertilized egg may be so influenced by various environmental factors that the embryos arising from such eggs are affected in a definite way. Similar effects from factors calculated to influence the sperm-cells are much more obscure. It is very difficult from the very nature of the factors involved to influence the sperm-cell in such a definite way that the influence will directly modify the offspring. Observations on this point have been numerous but few experiments under proper control have been made. In all experiments conducted for the purpose of influencing the male germ-cells, it is necessary to work through the animal body. Under these circumstances it is not always easy to determine with certainty whether the modifications resulting from various treatments are the direct result of the treatment on the sperm-cell or the secondary effects from the changes in the parental body itself. Among humans it is generally recognized that indulgence in alcoholic drinks by the male results in various defects in the offspring. In one observation Lippich studied ninety-seven children conceived during intoxication. Of this number all were defective except fourteen.[1] Among seven births

[1] Stockard, "Effect of Intoxicating the Male Parent," *American Naturalist*, vol. 47, p. 641; also *Journal of Heredity*, vol. 5, Feb. 1914.

from conceptions during drunkenness, Sullivan reports six as having died of convulsions and the seventh was stillborn. Chronic alcoholism has been found to change the structure of the testicular glands. The children of lead workers are known to be sometimes defective.

Stockard [1] has made some very interesting experiments which clearly show that the sperm may be so affected that the resulting offspring will be defective. As a result of mating normal female guinea pigs with males which were in a state of intoxication from inhaling alcohol fumes, many of the offspring were defective. "Out of 69 full term young, of which 54 were born alive, only 33 have survived and many of these are small and excitable animals, and although not treated themselves have since given rise to defective offspring in several cases where they have been mated with another."

If these results are confirmed, we must conclude that it is possible to modify the offspring by special treatment of the paternal parent. Some evidence is submitted by this investigation to show that the bad effects are not limited to the immediate offspring but are transmitted to subsequent generations.

49. Effect of lead poisoning on the male germ-cells as indicated by the offspring. — That the children of fathers who work in lead-manufacturing industries are often defective has been observed for a long time. Cole [2] and Bachhuber have reported results of feeding lead acetate to rabbits and fowls. The treatments were given to male parents alone and these were mated with normal females. Injury to the offspring from such treat-

---

[1] *Loc. cit.*
[2] Cole and Bachhuber, "Proceedings of the Society for Experimental Biology and Medicine," 1914, XII, pp. 24–29.

ments was frequent. The offspring of male rabbits treated with lead acetate were smaller and had distinctly lower vitality than the offspring of normal parents under similar conditions. The results from lead-poisoning of male fowls indicated that their offspring is of distinctly lower vitality than of offspring from normally healthy fowls.

The importance of these investigations to the practical breeder lies not in the fact that alcohol and other poisons may modify the male germ-cell and subsequently the offspring, but that the sperm-cell is capable of such reactions to environmental conditions that the progeny are profoundly changed or their development entirely prevented. If the offspring may be so modified through the sperm-cell by alcohol and lead acetate, then it is in all probability susceptible to other influences. It is a well-known fact that many conditions in ordinary breeding practice do modify the birth rate and the characters of the offspring. Certain feeds are known to have a more or less definite relation to the breeding powers. Investigations under proper control planned to test the influence of special feeds on the sperm-cell are lacking. At various times breeders have reported that alfalfa, clover, sugar and other materials when fed to breeding females have resulted in difficult conception or in weak offspring. Whether these feeding stuffs have an injurious effect on the male germ-cell is not known and cannot easily be determined in ordinary farm practice.

CHAPTER III

THE BREEDING SEASON

The arrival of puberty or the breeding age in the domestic animals does not mean that the breeding function is exercised continuously thereafter. In the case of all animals, domestic and wild, there exists a certain periodicity in the process of reproduction. The physiological activities which result in the propagation of young recur with a certain rhythm, and under normal conditions there is a definite period between the birth of young and the reappearance of reproductive activities.

This rhythm may be disturbed by certain external conditions, although in the higher animals it recurs with considerable regularity. The reproductive functions are one of the first to be affected by a marked change in the environment. Breeders have long recognized this fact. It has been observed that a stallion or bull imported from Europe to America often fails to breed well the first year. The same condition has been found to exist in the case of mares and cows.

50. Changed conditions. — Darwin has described how changes in the ordinary habits of animals may profoundly influence their reproductive functions. Animals in captivity rarely breed. Elephants, tigers, lions and many other species when confined fail to breed at all or breed

THE BREEDING SEASON

with great irregularity. Failure to breed under these conditions is not the result of a diseased condition of the generative organs; as Marshall has said, "It would seem probable that failure to breed among animals in a strange environment is due not, as has been suggested, to any toxic influence on the organs of generation, but to the same causes as those which restrict breeding in a state of nature to certain particular seasons, in that the sexual instinct can only be called into play in response to definite stimuli, the existence of which depends to a large extent upon appropriate seasonal and climatic changes."[1]

Among the domestic animals, the generative functions are more active in the spring season. This is true of the horse, the cow and the pig. Sheep breed more readily in the autumn.

How much the increased sexual activity of the domestic animals may be due to climate and how much to the change in the food supply, it is not easy to determine. Food itself as distinct from climate has a direct influence on the breeding powers of animals. The green succulent grass upon which the animals feed in the spring may be the efficient cause of increased sexual activity at that season. The ewes that exhibit an increased tendency to reproduction in the autumn may be stimulated in a similar manner by the general abundance of fresh feed which is characteristic of that season. It has long been the custom among skillful shepherds to provide fresh, succulent feed in abundance to ewes at the time of turning in the rams. This practice is called "flushing." The shepherds claim that this practice causes the ewes to come in heat more promptly and with greater regularity. It is also claimed that a larger number of lambs will be produced

[1] Marshall, "The Physiology of Reproduction," p. 5.

at lambing time as a result of this practice. It is undoubtedly true that the practice stimulates sexual activity and does cause the ewes to come in heat with greater regularity. It is probable that this result is due both to the character of the feed and its abundance.

The breeding season may be influenced by heredity. Certain breeds of sheep will breed readily at all seasons. The Dorset Horned breed comes in heat and breeds at all seasons, while most of the mutton and merino sheep breed readily only in the fall. Other conditions which are known to influence the breeding season in domestic animals are the kind of food, condition of the animal, age and breed.

Evvard [1] has shown that sows gaining in weight when bred produced larger litters than sows that were fed only a maintenance ration.

51. Phases of the breeding season. — The breeding season is divided into more or less distinct phases and these have been described and named by Heape [2] and Marshall as the procestrum, œstrum, metœstrum and diœstrum.

52. Procestrum. — The first part of the sexual season is occupied by the procestrum. This period is characterized by marked changes in the generative organs, the uterus becoming congested, while in the later stages there is often a flow of blood from the external opening of the vagina. The procestrum is the period often referred to by breeders as the time when an animal is "coming in heat" or coming in season.[3]

[1] Evvard, in "Report of American Breeders' Association," 191.
[2] Heape, *Quarterly Journal of Microscopical Science*, vol. **44**, p. 1.
[3] Marshall, "The Physiology of Reproduction," p. 36.

53. Œstrum. — The œstrum may be referred to as the heat proper and represents the time when the female will receive the male. It is during this period that the peculiar symptoms well known to practical breeders are exhibited. This phase is characterized by unusual activity on the part of the animal. Great restlessness, constant movement and often great mental excitement are observed in animals which are in heat. The genital organs become congested. The mammary glands in animals not suckling young increase in size. The external genitals, particularly the vulva, become swollen and red and mucous and bloody excretions flow from the generative organs. In many animals, there are frequent attempts at urination. The female sometimes utters loud cries or grunts, as in the sow. Cows, ewes and sows lose appetite, and get " off feed," often losing in weight. In all meat-producing animals, when the females are fattened for the markets, this is an economic loss to the feeder. In some sections, the larger feeders have spayed the heifers intended for fattening and thus prevented this loss. If the animal is bred and conception occurs, these periods of excitement are prevented, but pregnant fat animals are less valuable on the market and are always discriminated against by the buyer.

54. Metœstrum. — The œstrum is followed by a gradual subsidence of the symptoms which characterize this period, provided coition does not take place and pregnancy result. In the latter case, œstrum is followed by gestation. If the female is not bred to the male during œstrum, the sexual excitement of the period gradually passes away and the animal returns to a normal condition.

55. Diœstrum. — The diœstrum represents the time of rest for the generative system between the periods of sexual activity. This varies greatly in different animals. Again quoting Marshall, "In some animals, such as the dog, the metœstrous period is followed by a prolonged period of rest or anœstrum. In others, such as the rat or the rabbit, the metœstrum may be succeeded by only a short interval of quiescence. This short interval, which sometimes lasts for only a few days, is called the diœstrum. This in turn is followed by another prœstrous period, and so the cycle is repeated until the sexual season is over. Such a cycle (consisting of a succession of the four periods, prœstrum, œstrum, metœstrum, and diœstrum) is known as the diœstrous cycle. The number of diœstrous cycles in one sexual season depends upon the occurrence or nonoccurrence of successful coition during œstrus. Thus, if conception takes place during the first œstrous period of the season, there can be no repetition of the cycle, at any rate until after parturition. The cycle may then be repeated. If conception does not occur at any œstrus during the sexual season, the final metœstrous period is succeeded by a prolonged anœstrous or non-breeding period. This is eventually followed by another prœstrum, marking the commencement of a new sexual season. The complete cycle of events is called the œstrous cycle."[1]

56. Puberty. — The reproductive functions are not active in the young mammal during its very early existence. The nutritive system during the same period is characterized by unusual functional activity and a high degree of efficiency. The food consumed during the early existence of the mammal produces larger gains in live weight than the same food at any later period of its life.

The growth processes continue to function with great

[1] Marshall, "The Physiology of Reproduction," p. 37.

activity for a certain period and the reproductive organs gradually develop until at a certain age, or stage of development, the essential organs are matured and begin to develop perfect germ-cells. The stage of development when the ovaries of the female produce perfect eggs is called the period of puberty and represents the beginning of the breeding season. The arrival of puberty in the female is accompanied by the ripening of the first ovum or egg and the appearance of the œstrum or heat. The beginning of puberty in the female is marked by certain characteristic changes. The mammary glands increase in size, the general activities of the body are accelerated and the animal performs certain actions that are peculiar to the period of the œstrum or heat. The coming of puberty in the male is likewise associated with certain bodily changes which are well recognized by the practical breeder. In the stallion, the neck, and particularly the crest develops, and the forequarters generally are relatively better developed than the hindquarters. The most significant changes however, are physiological. The entire system assumes a state of greater activity. Not only the generative system is concerned in this change but all organs and functions of the body become more active. In the bull, the external and visible changes are an enlarging and thickening of the horns, and thickening and enlarging of the neck and crest. The increased activity of males is indicated by greater restlessness, irritability and the development of a pugnacious tendency. Bulls and stallions often become vicious and unmanageable and engage in deadly struggles for supremacy. These contests are also common among the males of wild animals. Such battles have been observed as common among stags and wild stallions.

57. Conditions influencing puberty. — The age at which puberty begins in the various breeds of the domestic animals is dependent upon the kind of animal, the breed, the food and general care of the young before puberty. Puberty in the mare comes between the ages of twelve and eighteen months. Under normal conditions, the stallion reaches the same stage at twelve or fifteen months. The cow and bull of the modern breeds of cattle under favorable conditions will reach this period at four to six months of age, but under ordinary conditions, not until they have attained the age of eight to twelve months. The ewe and ram of the mutton breeds will often arrive at the period of puberty at the age of five or six months, but generally will require longer. The sow and boar will reach puberty sometimes as early as three months of age, but generally at five to six months. The domestic hen has been known to lay eggs at the age of four-and-one-half months.

The beginning of puberty is greatly influenced by the nutrition of the young animal. This influence begins with the fœtus in utero. If the pregnant mother receives a generous supply of nutritious food during the period of gestation, the young will be better developed at birth. The reproductive system along with the other organs will have developed somewhat nearer to the stage of sexual completeness. If such generous nutrition is continued after birth, the young animal will reach the period of puberty at an earlier period than when fed on a poorer ration. At the Missouri Experiment Station, Eckles[1] has shown in an investigation comparing generous feeding with a lighter ration that the well-fed heifer on the average comes in heat 92 days earlier than those fed less generously.

[1] Eckles, "Dairy Cattle and Milk Production," p. 209.

58. The œstrum and lactation. — In many of the domestic mammals, the period of heat is influenced by lactation. Nursing the young may prevent entirely or retard the appearance of heat after parturition. Ewes seldom come in heat while suckling young. In the case of cows, the œstrum is retarded. There is much variation in the length of the time elapsing between the birth of the young and the first appearance of heat in the cow. It is probably about sixty days, although it may recur earlier or later than this period. The sow will generally come in heat three days after giving birth to a litter of pigs, but the œstrus period does not again occur until after the pigs are weaned. A prominent breeder of Duroc Jersey hogs, S. Y. Thornton, says, "In many cases, a sow that is in good condition will come in heat the third day after farrowing. I have bred them at that time but seldom knew one to get with pig if she was suckling, but one that has lost her pigs will invariably get with pig from the first period which is usually the third day after farrowing. A sow will often come in heat when her pigs are four to six weeks old if she has been well fed." The distinguished breeder of Berkshires, A. J. Lovejoy, says, "Sows often show signs of heat on the third day after farrowing, and again at eight weeks after farrowing, while suckling. We find that after weaning a litter, a sow will usually 'come in' in three to five days."

The heat period in the mare is very irregular. The "foal heat" occurs in seven to nine days after foaling. The œstrum recurs in most mares throughout the nursing period. But some mares do not come in heat during the time they are suckling foals. It is a well-known fact that there is a strong tendency in some mares to breed only once in two years. In some of the smaller mammals, pregnancy is common while the mothers are still suckling. This is true of the domestic rabbit, guinea pig and rat.[1]

59. Heat during pregnancy. — Animals do not normally come in heat during pregnancy. The fertilization of the egg of the female by the sperm-cell of the male sets in motion a series of physiological phenomena which react upon the ovaries in such a way as to cause a cessation of the heat periods. The ripening of eggs also does not normally occur during pregnancy. There are exceptions to this rule among certain mammals which seem to be otherwise entirely normal.

60. Superfœtation. — It sometimes happens that a pregnant mammal will not only come in heat and exhibit the various phases of the œstrum, but will ripen an egg and if bred to the male will conceive. Such an occurrence is given the name of superfœtation. This condition is somewhat rare, but has been observed more frequently among mares than other domestic animals. The reason for this is probably due to the fact that the records of breeding are more carefully kept for mares than for cows, sows or ewes. It is difficult even among mares to determine, when twins are born, whether these are the result of one mating or whether they may actually be of different ages. Such authentic cases as are known have been those observed in the mule-breeding districts of the United States. If the mare is first bred to a stallion and three or six weeks later to a jack, and twins are born, one a mule and the other a horse, there can be no doubt that these colts are of different ages. Such a result could only happen in mares which come in heat and ripen an egg during pregnancy.

[1] Heape, *Quarterly Journal of Microscopic Science*, vol. **44**, p. 43.

61. Examples of superfœtation. — The literature of animal-breeding is singularly lacking in records of authentic cases of superfœtation. The writer has through many years collected evidence of such cases in the mule-breeding districts of Missouri, and a few of these are recorded here:

"W. E. Carmichael of Shelbyville, Missouri, bred a mare to a stallion and thirty days later to a jack. At

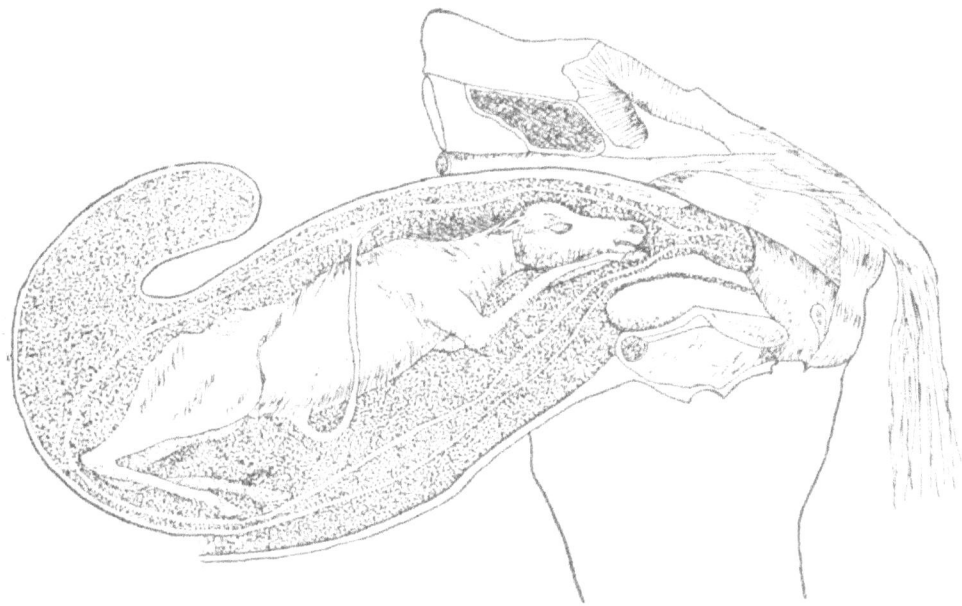

Fig. 11. — Normal and usual anterior presentation in mare.

the end of the normal period of gestation the mare gave birth to twins, one a mule and the other a horse colt. They were both dead at birth."[1]

A nine-year-old mare belonging to Eugene Rhodes of Fairfax, Missouri, was bred to a stallion in May. In August she showed unmistakable signs of heat and was mated with a jack. In the month of January following she gave birth to a perfectly formed mule colt and a

[1] Mumford, "Amer. Cyclopedia of Agriculture," vol. III, p. 31.

horse colt, the latter being considerably shriveled in appearance.

A case is reported by F. K. McGinnis of a Texas bred mare belonging to Mr. Carmack. This mare was bred several times to a stallion during a period of six weeks. She continued to come in heat and the owner, concluding that she was sterile with the stallion, bred her to a jack. She was mated with the jack a number of times during a period of two weeks. Her owner, finally despairing of ever getting the animal in foal, turned her out. At the end of eleven months from the time the mare was first bred to the stallion she dropped twin colts, one a horse and the other a mule. Both were dead at birth.

An interesting example of superfœtation is recorded by J. F. White of Whitesville, Missouri, who bred a four-year-old mare to a saddle stallion on April 25, 1909. She came in heat again and was bred to a jack May 29, 1909. She was again in heat June 12, 1909, and was bred at this period to the saddle stallion first mentioned. On May 11, 1910, the mare gave birth to perfectly formed twin colts. One of these was a mule and the other a horse colt. The mule died from illness at the age of three weeks. The mule colt was a male. The horse colt was a mare and developed into a perfect colt. These were the mare's first colts.

A draft mare belonging to J. C. Spies of Newark, Missouri, was bred to a jack. Later she was turned in a lot with her own two-year-old stud colt. The following spring she foaled a mule and a horse colt. The mule died at five days old. The horse colt lived and made a good horse.

Cases of superfœtation where the same animal is the

sire of both twins are probably much more common than is generally believed. In such cases it is difficult to determine whether the twins foaled at the same time are the result of the fertilization of two eggs ripened at one and the same heat period, or whether the eggs have been ripened at different periods some distance apart.

A case of this kind has been reported by J. A. Finley of Troy, Missouri. In this case the mare was bred to a jack, and twenty-one days later she was found in heat again and was again bred to the same jack. At the end of a few months she aborted, losing twin mule colts. One of these was much better developed than the other. It seemed clear that the two colts were from different periods of heat,— in this instance, twenty-one days apart.[1]

The following examples are probably to be regarded as cases of superfœtation:— A sow belonging to O. Young of Hopkins, Missouri, gave birth to three pigs. The mother nursed them for four or five days, when she weaned them. Three weeks later the sow gave birth to eight pigs, six of which lived and became thrifty young hogs.

A young ewe owned by A. Cassity of Linneus, Missouri, gave birth to twin lambs on February 13th. About six weeks later on March 30th, she gave birth to a third lamb. See Plate I.

62. Recurrence and duration of the œstrum. — The occurrence and duration of heat are influenced by age, species, food supply, season, heredity and other conditions. The heat period persists in the domestic animals for one to fifteen days.[2] The duration of the heat period is

[1] From letter to Geo. F. Nardin, dated June 24, 1912.
[2] Hill, "Bovine Medicine and Surgery."

shortest in the cow and sheep, being in these species usually from twelve to twenty-four hours.[1] Mares come in heat seven or nine days after foaling. Bred at this time the mare is more certain to conceive. Dimon,[2] an experienced horseman, says that there is no regular period for the return of heat in mares. If mares are well fed they may come in heat at any season, but are more generally in heat in the spring and fall.[3] If the mare fails to become pregnant when bred on the ninth day after foaling, she will usually be in heat twenty-one days thereafter. Heat persists for five days and recurs every twenty-one days.[4] Heat recurs in the cow three or four weeks after parturition and recurs every twenty-one days. The heat period recurs in sheep from seventeen to twenty-five days.

In the sow the heat period will be observed three days after delivery and usually not again until the pigs are weaned. Heat again recurs three or four days after weaning the pigs and every twenty-one days thereafter.

63. Effect of ration on recurrence of œstrum. — The recurrence of the œstrum after delivery of the young is often influenced by the character of the ration. At the Wisconsin[5] Experiment Station, Hart, McCollum, Steenbock and Humphrey found that pregnant cows fed on an exclusive corn ration came in heat in four to six weeks after the first calf. When the cows were fed a ration

---

[1] Weber, "Untersuchung über die Brunst des Rindes," Arch. f. Wissensch. u. Prakt. Tierheilk., 37 Bd., 1911.

[2] Dimon, "American Horses and Horse Breeding."

[3] Reynolds, "The Breeding and Management of Draught Horses."

[4] Curtis, "Cattle, Horses, Sheep and Swine."

[5] Wisconsin Exp. Station, Bul. No. 17.

made up exclusively of wheat and its products, the first appearance of heat was from ten to eighteen weeks after calving. Fresh pasture also will often cause animals to come in heat earlier after delivery, and the œstrum will recur with greater regularity.

CHAPTER IV

GESTATION AND LACTATION

FROM the time the egg is fertilized until the young animal is able to live an independent life covers a period which is of the greatest importance to the growth and development of the individual and the mother. During this time all the physiological activities which are concerned with growth are at maximum efficiency; at no other period in the life of the animal is growth so rapid. Not only is the rate of growth very rapid, but the food is utilized much more economically. The stage of development which takes place in the uterus of the mother is the period of gestation. The period of lactation is the time during which the mammalian animal elaborates milk.

GESTATION

64. Indications of pregnancy. — If the animal comes normally in heat and is bred to the male during the heat period, conception occurs in the natural course of events and pregnancy begins as a result of successful conception. Significant physiological changes occur which are recognized by the breeder as evidences of pregnancy. Pregnant animals do not normally come in heat. The chief evidence, therefore, that an animal is "in foal," "in calf," "in pig," or "in lamb" is the cessation of the periodic appearance of the symptoms accompanying the period of the œstrum. If after breeding the mare does not come in heat for thirty days, she is probably safe in foal. The heat period in the mare persists for several days, and therefore a reappearance of evidences of heat shortly after breeding should not be regarded as significant. It is pointed out elsewhere that mares may sometimes come in heat and conceive again (see superfœtation), even though already pregnant. It is also true that some mares will persistently refuse the horse, even though not pregnant. In the cow a period of sexual quiescence for three weeks following her breeding with the bull is good evidence that she is safely " settled " and in due time will give birth to offspring.

The beginning of pregnancy in an animal is often accompanied by a marked change of temperament. A nervous, excitable mare may become more gentle and docile. It is also true that some mares which when not pregnant are quiet and gentle with other horses become cross during pregnancy and evince a desire to fight other horses. This tendency increases as pregnancy advances. Following conception the pregnant animal shows a tendency to lay on fat much more rapidly. Feeders sometimes take advantage of this tendency to finish heifers and sows rapidly for the market, but such a practice is to be condemned, as the meat from pregnant animals is less desirable for human food. As pregnancy advances, the abdomen becomes larger at the sides and below and the flank falls in. The loins become depressed, owing to the sinking of the spine due to the increased weight of the abdomen. This depression of the loins gives the croup bones the appearance of rising. The udder of the pregnant animal is not materially changed during the initial stages of gestation, but during the later stages this organ gradually expands and the teats become larger. A short time before parturition the udder becomes greatly swollen and a waxy substance exudes from the ends of the teats. In the cow the developing fœtus may be observed externally after the fifth month of pregnancy. If the cow is permitted to take a drink of very cold water, the movements of the young calf may be felt by pressing the hand against the flank just in front of the stifle. In the mare the same movements of the unborn foal may be observed from the seventh to the eighth month in a similar manner by pressing the hand firmly against the flank in front of the left stifle.

65. Physical examination for pregnancy. — The existence of pregnancy may be determined with considerable accuracy by examination through the rectum. The method of making this examination is described in such an admirable manner by Law [1] that it is quoted entire: "Examination of the uterus with the oiled hand introduced into the rectum is still more satisfactory, and if cautiously conducted no more dangerous. The rectum must be first emptied and then the hand carried forward until it reaches the front edge of the pelvic bones below, and pressed downward to ascertain the size and outline of the womb. In the unimpregnated state the vagina and womb can be felt as a single rounded tube, dividing in front to two smaller tubes (the horns of the womb). In the pregnant mare not only the body of the womb is enlarged, but still more so one of the horns (right or left), and on compression the latter is found to contain a hard, nodular body, floating in a liquid, which in the latter half of gestation may be stimulated by gentle pressure to manifest spontaneous movements. By this method the presence of the fœtus may be determined as early as the

---

[1] Law, "Diseases of the Horse," U. S. Department of Agriculture, p. 155.

third month. If the complete natural outline of the virgin womb cannot be made out, careful examination should always be made on the right and left side for the enlarged horn and its living contents. Should there still be difficulty, the mare should be placed on an inclined plane, with her hind parts lowest, and two assistants, standing on opposite sides of the body, should raise the lower part of the abdomen by a sheet passed beneath it. Finally the ear or stethoscope applied on the wall of the abdomen in front of the stifle may detect the beating of the foetal heart (one hundred and twenty-five per minute) and a blowing sound (the uterine sough), much less rapid and corresponding to the number of the pulse of the dam. It is heard most satisfactorily after the sixth or eighth month and in the absence of active rumbling of the bowels of the dam."

66. The period of gestation. — The period of development from the fertilization of the egg by the sperm-cell until the birth of the fully developed offspring capable of independent existence outside the body of the mother is known as the period of gestation. Among oviparous animals it is the period of incubation. This period varies greatly as between different species, but under normal conditions is fairly uniform in animals belonging to the same species. The normal period of gestation has in general a more or less definite relation to the size of the animal. The length of gestation in animals as reported by various authors [1] is as follows:

---

[1] Nathusius, "Zool. Garten Jahrg.," 3, 1862.

Heape, "The Sexual Season," *Quarterly Journal of Microscopical Science*, vol. 44, 1900.

Ewart, "The Development of the Horse," *Quarterly Journal of Microscopical Science*.

Wortley Axe, "The Mare and the Foal," *Journal of the Royal Agricultural Society*, 3d series, vol. IX, 1898.

Elephant	20 to 23 months
Giraffe	14 months
Dromedary	12 months
Buffalo	10 to 12 months
Camel	13 months
Jennet	12 months
Seal	11 to 12 months
Mare	11 to 12 months
Zebra and Celtic Pony	334 to 338 days
Prjewalsky's Horse	356 to 359 days
Cow	9 months
Bear	6 months
Reindeer	8 months
Monkeys	7 months
Sheep and Goat	21 to 22 weeks
Sow	4 months
Lion	3½ months
Dog	59 to 63 days
Fox and Wolf	63 to 63 days
Guinea Pig	61 days
Cat	63 days
Polecat	40 days
Rabbit	30 days
Squirrel and Rat	28 days
Mice	21 days

Small breeds require a shorter time than larger breeds, but this influence may be overcome in breeds which have long been selected for early maturity. According to Youatt,[1] all domestic animals are subject to considerable variation in the length of gestation both above and below the normal period. Tessier[2] has observed a large number of pregnant females among the domestic animals and has reported marked variations in the time required for complete development of the fœtus and final expulsion from the uterus. This authority has reported on 582 mares in which the period of gestation ranged from

[1] Youatt, "Cattle," p. 521.
[2] Tessier, "Recherches sur la Durée de la Gestation," Mem. de l'Acad. des Science, Paris, 1817.

287 to 419 days. Among 1131 cows the length of the period varied from 240 to 321 days. The minimum gestation period among 912 sheep was 146 days and the maximum 161 days. It is of interest to note, however, that among 676 ewes the period varied from 150 to 154 days only.

The Earl of Spencer[1] found the period to vary from 220 to 313 days in cows. The following table includes the observations of a number of investigators and is useful as indicating the somewhat wide variations which may occur in the gestation period of the domestic animals:

PERIOD OF GESTATION IN DOMESTIC ANIMALS[2]

	NUMBER OF CASES	MAXIMUM DAYS	MINIMUM DAYS	AUTHORITY
Mares	582	419	287	Tessier
Mares	25	367	324	Gayot
Cows	575	299	240	Tessier
Cows	764	285	220	Spencer
Cows	50	291	268	Allen
Cows	98	299	276	Bement
Cows	182	296	280	Wing
Ewes	912	161	146	Tessier
Ewes	420	156	143	Magne
Sows	25	123	109	Tessier
Sows	10	116	101	Fox

[1] *Journal of the Royal Agricultural Society*, vol. I, pp. 166, 167.

[2] See Tessier, *Journal of the Royal Agricultural Society*, vol. I, pp. 166, 167.

Franck-Albrecht-Goring, " Die Trächtigkeitsdauer," Thierartzliche Geburtshülfe, vol. 4, 1901.

Wing, Cornell Experiment Station, Bul. 162.

Smith, "Physiology of the Domestic Animals."

Allen, "American Cattle," p. 259.

Miles, "Stock Breeding," 1907, p. 400 *et seq.*

The period of gestation in the domestic jennet is 370 days. Careful records kept by Kalo Monsees[1] on the large jack- and jennett-breeding farm of Monsees and Sons at Smithton, Missouri, indicates that the maximum gestation period for a living colt was 13 months and 16 days. The minimum period was 11 months and 15 days. The average period was 370 days. These figures may be taken as reliable since they cover a large number of animals of varying ages and sizes for several years and represent the progeny of different jacks.

67. Causes of variation in length of gestation period. — The causes of variations in the time required for the development of the young in the uterus are not always clearly apparent. It is probable that prolonged gestation may sometimes be due to the mother suckling young during pregnancy, resulting in providing an insufficient supply of food to the developing fœtus.[2] Tessier concludes that the length of the period of gestation does not depend upon age, constitution of the female, diet, breed or season. Nathusius[3] and others have found that comparing races and breeds belonging to the same species, those which have been selected for their early maturing qualities have a shorter gestation period. Darwin[4] has given some evidence of this in connection with the grading up of Merino sheep bred to the earlier maturing Southdown. The length of gestation is given as follows:

[1] Kalo Monsees, "Breeding Records of Monsees and Sons Jack and Jennett Breeding Farm."
[2] Pinard, "Gestation," Richets Dictionnaire de Physiologie, vol. 7, Paris, 1905.
[3] *Loc. cit.*
[4] Darwin, "Animals and Plants under Domestication," vol. 1, p. 123.

Merinos	150.3 days
Southdowns	144.2 days
Half blood Merino and Southdown	146.3 days
Three quarter blood Southdown	145.5 days
Seven eighths blood Southdown	144.2 days

Some doubt has been expressed as to the authenticity of the maximum period of gestation in mares. The editor of the Breeder's Gazette commenting on this fact says, "Mares are usually credited with pregnancy lasting eleven months. When they run twelve months we prefer to believe that the date has not been properly kept. We believe that forty-eight weeks, seven days to the week, or 326 days is about the average duration of pregnancy in a mare."[1] Referring to this statement M. W. Johnson[2] of Illinois writes that he has been in the breeding business for fourteen years and has complete records on the breeding of 5000 mares. He reports one mare as having carried her foal for twelve months and sixteen days and another for twelve months and eighteen days. Both foals were deformed. Another mare gave birth to a perfectly healthy foal at the end of one year and eighteen days while another carried her foal exactly thirteen months and dropped a healthy foal. A Percheron mare belonging to H. F. Sperry[3] dropped a mule colt twelve months and two days after breeding. William Lokings of South Dakota owned a pony mare which was twice bred and gave birth to a foal twelve months and twenty-five days after the last service. There is a popular belief that male offspring are carried longer than female. Bement[4] found that the average length of gestation for male calves

---

[1] *Breeder's Gazette*, May 15, 1907.
[2] *Ibid.*, June 19, 1907.
[3] *Ibid.*, June 26, 1907.
[4] *The Cultivator*, 1845, p. 207.

was 288 days and for females 283 days, but in 1839 the female calves were carried longer than the males. Earl Spencer [1] also believed there was some relation between sex and the length of gestation. M. Magne [2] found the period of gestation longer for ewe lambs than for ram lambs. C. U. Connellee [3] of Texas finds no relation between the sex of foals and the time required for gestation. The evidence available is insufficient to justify the belief that male offspring are carried longer than females.

68. Incubation. — The period of incubation in fowls represents in oviparous animals the phenomenon of gestation in mammals. The length of the period of incubation among domestic birds is given by Miles [4] as follows:

"Turkey, twenty-six to thirty days; guinea hen, twenty-five to twenty-six days; pea-hen, twenty-eight to thirty days; ducks, twenty-five to thirty-two days; geese, twenty-seven to thirty-three days; hens, nineteen to twenty-four days, or an average of twenty-one; pigeons, sixteen to twenty days; canary-birds, thirteen to fourteen days. Mr. Wright remarks that 'cold weather, or a prevailing east wind, will lengthen the time a day or more, while warm weather and an attentive sitter will hasten it; stale eggs also hatch later than fresh.'"

The smaller breeds, like bantams, hatch in nineteen or twenty days, while the heavier breeds may require as long as twenty-two days for complete incubation. When eggs are artificially incubated, it has been found that a higher temperature combined with favorable moisture conditions will shorten the period.

[1] Spencer, *Journal of the Royal Agricultural Society*, vol. 1, p. 168. [2] *Loc. cit.*
[3] *Breeder's Gazette*, June 26, 1907.
[4] Miles, "Stock Breeding," p. 401.

69. Parturition. — In mammalian animals at the end of the period of gestation and in the normal course of events, the fully developed fœtus is expelled from the uterus. This phenomenon is known as parturition. The beginnings of parturition are accompanied by a series of rhythmic contractions (labor pains) of the uterus which eventually result in the birth of the offspring. These contractions are at first partially controlled by the will, but later are entirely involuntary. That the muscular movements of the uterus are not controlled entirely by the central nervous system is shown by the researches of Kehrer,[1] Helme and others. These investigators found that a healthy uterus may rhythmically contract when separated from the body if it is maintained at body temperature without important variations. The powerful muscular contractions of the uterus of mammals are characterized by the mechanical stretching of the bag of membranes by severe contraction of the longitudinal muscle fibers and the relaxation of the circular fibers of the cervix. The contraction of the uterus and the relaxation of the cervix causes the bag of membranes to act as a fluid wedge still further extending the neck of the womb. These phenomena are followed by the head and fore legs of the young animal, the rhythmic contractions become more frequent and more powerfully exerted, and these are supplemented by the abdominal muscles in the final stages of parturition. The immediate inciting causes of parturition are not well known. Various explanations have been attempted. Spiegelberg[2] has suggested that the fœtus secretes a substance which

---

[1] Marshall, "The Physiology of Reproduction," p. 527.

[2] Spiegelberg, "Die Dauer der Geburt," Lehrbuch der Geburtshülfe, vol. II, 1891.

finally entering the maternal blood reaches the nerve centers and through them acts on the uterine nerves in the spinal cord. Beard[1] and others have held that there is an intimate relation between the œstrus cycle and parturition. It is known that abortion is more apt to occur at the time of the periodical return of heat. But may it not be true that the period of gestation is itself governed by a certain rhythmic law or periodicity similar to that which brings about the heat period? Other possible explanations are that the placenta begins to decay or atrophy at the end of a given period and thus loses its hold on the uterine walls, and the waste products thus developed may furnish the real stimulus which results in parturition.[2] The death of the fœtus prematurely will generally bring on expulsive movements of the uterus and speedily relieve the uterus of its dead burden.

It is probable that the causes of parturition are very complex and that a combination of the above, together with other causes, may bring about eventually the successful birth of the young mammal after the normal period of existence in the uterus.[3]

70. Normal parturition of the domestic animals. — Under ordinary conditions and in due course of time, the young of the domestic animals are born without difficulty. When difficult parturition occurs, it may be due to mal-presentation of the fœtus, malformation of the fœtus, malformation of the mother, or disease of the mother preventing the normal expulsion of the fœtus.

[1] Beard, "The Span of Gestation and the Cause of Birth," Jena, 1887.
[2] Williams, "Obstetrics," London, 1904.
[3] *See also* Marshall, *loc. cit.*

71. Mal-presentations. — In order that the offspring may be successfully expelled from the generative organs of the mother, it is necessary that it should approach the neck of the uterus in a certain form or " presentation." In all the mammalian domestic animals the normal presentation is one in which the fore legs are extended forward with the nose also extended forward and lying between the knees. (See Fig. 11.) In case of twins,

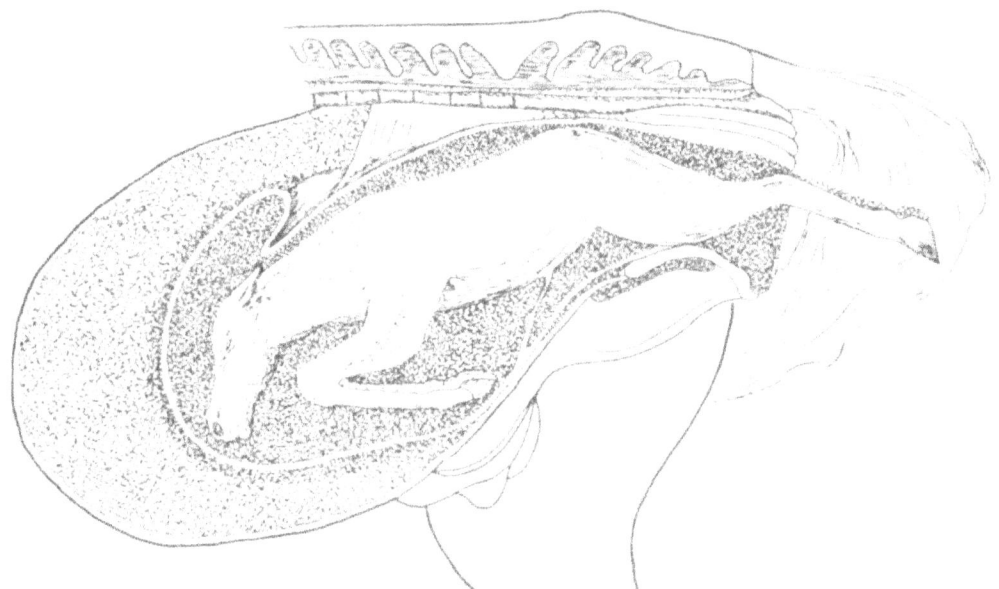

Fig. 12. — Parturition in mare. Posterior presentation.

the second is generally presented with the hind feet first. (See Fig. 12.) Any other presentation than the two here described is abnormal, and the birth of young under such conditions is difficult or impossible.

As most cases of difficult parturition are due to abnormal presentations, it is important that the more common mal-presentations should be mentioned. The more common and difficult mal-presentations are: — head normal, but fore legs bent back at the knee (Fig. 13); head normal, but fore legs bent back from the shoulders and entirely under the body; fore feet normal, but head bent to one side or downward; fore legs normal, but the head bent backward and upward; all four feet presented; thigh and croup presented first, with hind legs bent under body (Fig. 14); the back first presented, with fore and hind legs extending backward toward the uterus (Fig. 15).

72. Normal presentations. — In the early stages of parturition, it is desirable to determine whether the foetus is normally in position to be expelled with least difficulty. As already described, the normal presentation is head and fore legs forward, or in some cases (twins) the hind legs forward. In making the examination for the purpose of determining whether the foetus is in proper position, the hand and arm should be thoroughly cleansed and oiled with vaseline, and inserted into the vagina and an examination made. The manipulation should be conducted with extreme gentleness and under such conditions as shall not excite the pregnant animal. Advantage should be taken of the lulls in the labor pains to make the examinations. While the muscular contractions are on in full force, little can

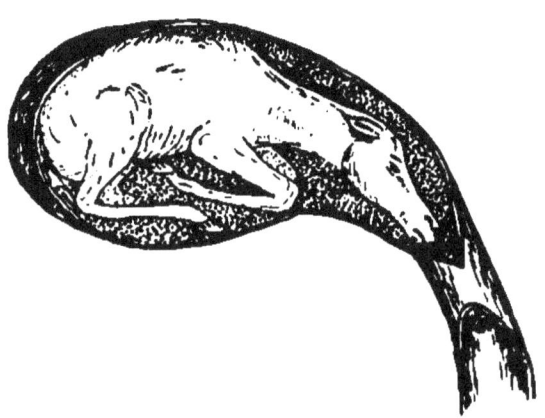

Fig. 13. — Abnormal anterior presentation.

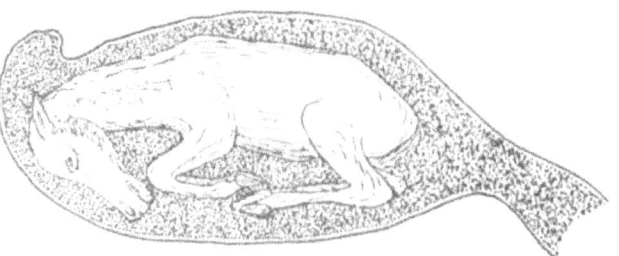

Fig. 14. — Abnormal posterior presentation.

be accomplished. If the fœtus is found to be normally presented, it is always wise to let nature take her course, and in most cases birth ensues without assistance from the attendant.

73. Treatment for mal-presentation. — If the examination reveals the fact that the fœtus is not normally presented, an effort should be made to readjust the

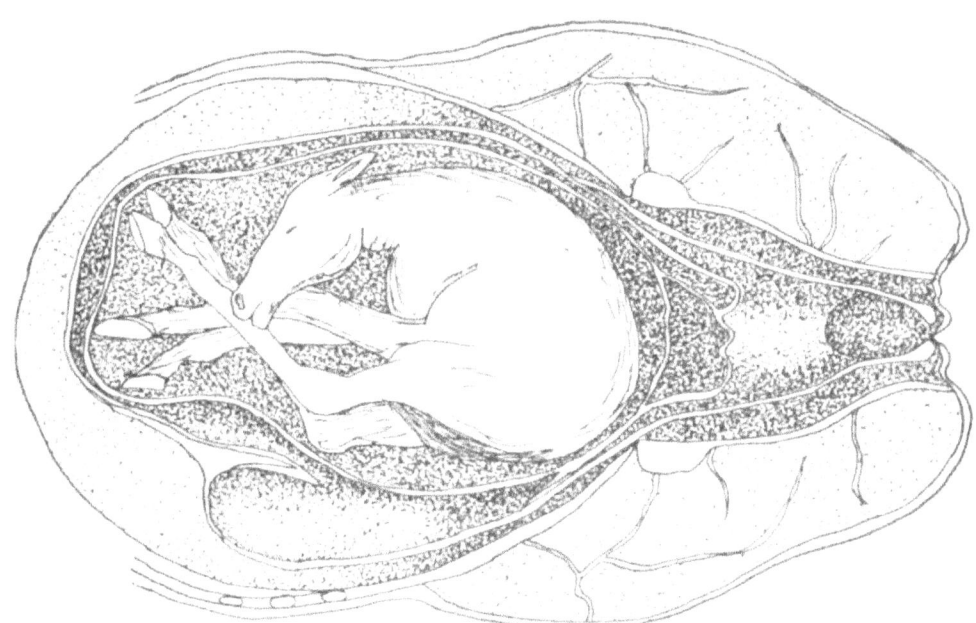

Fig. 15. — Abnormal transverse presentation.

unborn animal so that it will be normally presented. In the case of valuable breeding animals, it is generally best to secure at once the services of a skilled veterinarian. In attempting to make the readjustment of the fœtus, the same care and gentleness should be exercised as in the initial examination. In order that the mal-presentation may be successfully adjusted and the fore legs and head properly brought forward into the cervix and vagina, it will be necessary to push the fœtus back into the uterus where there will be sufficient room for the manipulation.

If the labor pains have already proceeded for some time, it may at first be found somewhat difficult to return the fœtus to the uterus. In all cases it will be useless to attempt to push back the unborn animal during the severe labor pains. But as the resting period and consequent relaxation follow each severe contraction, the fœtus may be gradually pushed back. It is sometimes helpful partially to suspend the hind quarters of the pregnant mother by roping the feet and hoisting the hind quarters so that they will be somewhat higher than the forequarters, and in this position it is generally easier to accomplish the return and readjustment. In most cases in which the fœtus and mother are in normal health and condition, the fœtus may be expelled without great difficulty after readjustment, that is, provided the mother has not become too severely exhausted by long-continued labor. In such case it is necessary to render aid by supplementing the mother in her efforts to expel the fœtus.[1]

LACTATION

The young of most mammalian animals are born into the world in an immature and often quite helpless condition. In most species the newborn animal is unfit to live and thrive independently. In particular the nutritive functions of the very young mammal are not developed to a point where the individual can immediately exist on the food consumed by the mature parent. To provide nourishment for the very young animal, all mammals are provided with mammary glands which secrete milk.

---

[1] See Law, "Diseases of Cattle," 1908, and "Diseases of the Horse," 1907, U. S. Department of Agriculture.

74. The mammary glands. — All mammals are supplied with milk-secreting glands. These glands are present in both sexes, but are rudimentary in the male. In the female the glands are large and are stimulated into active functioning by the exercise of the reproductive organs. In the immature female before the arrival of puberty, these glands are small and inconspicuous. With the first appearance of puberty accompanied by the œstrum, the glands increase in size.

The number of glands present varies with the species. In some animals which normally give birth to one offspring at a time, as in man, there are two glands (mammæ). In the cow, however, normally producing one at a birth, the normal number of mammæ is four. In cats, dogs, rabbits and swine, species producing from four to twenty young at one time, there are normally present several pairs of mammæ. Wentworth found the number of mammæ in swine to vary from nine to fourteen.[1] There is, therefore, a rather definite relation between the number of mammæ and the normal number of young produced at each birth. This relation does not seem to be important as an index of fertility in any particular species.

75. The duration of lactation. — Among wild forms the continuance of lactation varies widely in different species. In general the period of lactation ends when the young animal has developed to a point where it can live independently and secure its nourishment in the same way as the mature individual. In the case of the domestic cow, the milking function has been developed and stimulated under domestication to a point where the lactation period may persist from one calving period to the next without

[1] Wentworth, "Inheritance of Mammæ in Duroc-Jersey Swine," *American Naturalist*, vol. 47, 1913.

intermission. The more important conditions which influence the duration of lactation and the amount of milk produced among the domestic animals are food supply, habit, heredity, exercise, climate and nervous excitement.

76. The food supply. — Every mature domestic animal requires a certain minimum amount of food to maintain the ordinary bodily functions. This is called the food of maintenance. The function of milk-giving must be regarded as an additional requirement. The animal, therefore, that is producing milk must consume and assimilate larger quantities of food than one that is dry or not producing milk. It follows that the greater the amount of milk produced, the larger the demands of the lactating animal for food. In an ordinary cow in full milk, the food of maintenance may represent sixty per cent of all the food eaten. In this case forty per cent of the ration is available for milk production. In a heavy-producing cow, the food of maintenance may represent only forty per cent of the whole. In the latter case, as much as sixty per cent of the ration may be utilized for milk production. The greater economy of production in the case of the heavy-producing cows is at once apparent. If an insufficient ration is fed to any cow in full milk, the first effect will be to cut down the milk flow.

77. Habit. — The duration of the lactation period is materially influenced by the habit of the cow as determined by man. The lactation period of a cow, which is normally ten months, may be shortened by careless milking or insufficient feeding. The milking function may be so stimulated by careful and thorough milking and intelligent feeding that the daily quantity of milk may be increased and the period of lactation lengthened.

78. Heredity. — The capacity to give milk in abundance

is hereditary. The present highly productive dairy breeds owe their greater ability to produce large quantities of milk to their inheritance of this quality from their ancestors. In the same breed certain families are known to possess the quality of large capacity for milk production to a higher degree than the average of the breed. It is also true that the duration of the period of lactation is influenced by heredity.

79. Exercise. — An excessive amount of muscular exertion of any kind must be regarded as unfavorable to the maximum production of milk. Cows that are required to travel long distances over sparse pastures in order to secure sufficient food for their needs cannot produce their maximum quantity of milk. In many European countries, cows are generally employed as draft animals in the ordinary farm operations of plowing, harrowing, reaping and other work. Investigations in Germany have shown that so long as cows are employed at moderate work the milk flow is not decreased. Whenever cows were compelled to work at heavy labor and for long hours, invariably the flow of milk was decreased.

80. Climate. — Exposure to extreme dry cold or to cold driving storms will have the effect of decreasing the normal milk yield of a herd of cows. In general, a lack of adequate shelter in a cold and humid climate may seriously interfere with the highest development of the milking functions. A herd of cows will produce more milk in winter if provided with water which has been slightly warmed.

81. Unusual lactation. — In general, lactation is closely associated with reproduction. Pregnancy followed by parturition normally precedes the secretion of milk in the mammary glands. Males and sterile females possess rudimentary mammæ but seldom secrete milk, although a number of cases are on record where males have been known to secrete milk. The hybrid mare mule in rare cases has been known to secrete milk through the excitation of the mammary glands. In the cases observed, the mammary glands have generally been stimulated to secrete through the persistent suckling of some young mule colt which may have been running in the same pasture or lot. It is rare that the mammary glands of a mare mule will develop to the point of secreting milk in the absence of some such stimulation.

The author has discovered one mare mule which has secreted milk without any such stimulus. This mare mule belonging to L. O. Swarner of Boonville, Missouri, when first observed had been giving milk for some weeks. The owner in a letter to the author says: "I have a mule that has been giving milk the same as a brood mare does when suckling a colt. She has been giving milk for about five weeks. The mule is still giving milk, as much as a quart at a time. (See Plate III, upper.) The milk is pure white and streams from her udder." This mare mule came in heat regularly and was bred several times but failed to become pregnant. The milk was analyzed by the Missouri Experiment Station. The analysis is shown in the table in comparison with cow's and human milk:

Composition of Milk from Mare Mule Compared with Cow's and Human Milk

	Mare Mule "Beck"	Cow's Milk	Human Milk
Water	90.441%	87.17%	87.41%
Total Solids	9.559	12.83	12.59
Ash	0.400	0.71	.31
Fat	1.450	3.64	3.78
Protein	2.463	3.55	2.29
Sugar	5.792	4.88	6.21

CHAPTER V

FERTILITY

The larger number of the breeders of domestic animals are engaged merely in the multiplication of animals. They are not primarily striving for the improvement of the species. To all these, the ability of an animal to produce young in abundance is of fundamental importance. To the relatively small class of breeders who are successful in really improving the desirable characteristics of existing breeds, the quality of fertility is likewise of primary importance. When the breeder has succeeded in developing a highly improved strain, it becomes important to secure as many offspring possessing the new and desirable qualities as possible.

Fertility may be defined as the ability of an animal to produce young in abundance. This quality depends upon the number of young born at one time, the frequency of the recurrence of the œstrum, the duration of the period of gestation, and the length of the period of life during which reproduction occurs. All of the above conditions are affected by many circumstances, some external, others internal and inherent in the individual and the species. Many of the circumstances influencing fertility can be directly or indirectly controlled by man, others are beyond his control.

82. The number of young at a birth. — There is very great variation among animals in respect to the number of young born at one birth. This difference is very marked, as between different species, for example, as between the sow and the ewe. A similar difference is likewise to be observed between individuals and families belonging to the same species. In a general way, the number of young carried in the uterus at one time seems to depend upon the size of the animal. Thus the elephant, rhinoceros, hippopotamus, giraffe, bison, domestic mare and cow, produce one young at a birth. The goat and sheep, while normally producing one offspring at a time, may frequently produce twins and triplets or even a larger number at one birth.

The hare and rabbit are much smaller than the sheep and are very much more prolific. The wild rabbit produces from four to eight in a litter. The domestic rabbit is more prolific than the wild, often giving birth to eight or ten at one time.

The lion and tiger in a wild state give birth to two to three cubs, while the domestic cat will sometimes produce as many as nine. The size of the litter in the fox is from four to eight. The number of young in the litter of the domestic dog varies from five to twelve.

The domestic sow is an exception to the general rule that smaller species are more prolific. The sow is the most prolific of the more important domestic animals, though not the smallest. The wild sow's litter numbers four or five. The domestic sow produces from seven to twelve, and much larger litters are common.

The smaller rodents, like the rat and mouse, are characterized by large litters. The rat regularly gives birth to twelve or fifteen young at a time and has been known to produce twenty at one birth. The mouse is equally prolific.

It does not seem to be generally true among domestic animals that the smaller breeds are more fertile than the larger breeds of the same species. Among sheep, the American Merino and Southdown are relatively small breeds, but are less prolific than the larger Shropshire, Cotswold or Lincoln. The smaller breeds of swine, like the Essex, Cheshire and Small Yorkshire, are in general less prolific than the larger Berkshire, Duroc-Jersey and Large Yorkshire. In this instance, it is probable that selection by man has intervened to counteract the general law that the smaller animals are more fecund.

83. Period of gestation and fertility. — The length of the period of gestation seems to be some index of the number of young produced at a birth. In all animals requiring a longer period of gestation than six months, one is the normal number of young at a birth.[1] Among the animals in which the period is less than six months are the hog, sheep, goat, rabbit, dog, cat, rat and mouse. In all of these, except the sheep and goat, the normal number is greater than one. The sheep and goat under domestication have so increased in fertility, in the cases of some breeds at least, that fecundity is much greater than in the horse or cow, where single births are the rule. Milch goats usually drop twins, and triplets are not rare.[2]

The Border Leicester breed in England has produced 150 to 160 per cent of lambs under ordinary conditions. When the same breed has been specially fed before breeding, the number of lambs has been increased to 200 per cent.

[1] Marshall, "The Physiology of Reproduction."
[2] Clos, vol. III, p. 410, "Cyclopedia of American Agriculture."

84. Fertility and the frequency of the recurrence of the œstrum. — The fertility of animals is also dependent upon the frequency of the recurrence of the œstrum. As already described, the recurrence of heat depends upon a number of conditions, chief of which are pregnancy and lactation. Certain of the domestic animals do not normally come in heat while suckling young, but to this there are many exceptions. Animals having a short period of gestation rarely or never come in heat while suckling young.

85. Fertility and gestation. — Animals having a long period of gestation are less fertile than animals requiring a shorter time for the development of the young in the uterus.

The exceptional fertility of the domestic sow is due not only to the large number of young born at a time, but to the further fact that the period of gestation is only four months. A sow may thus easily produce two litters a year. The mare produces but one offspring at a time, and the period of gestation is from eleven to twelve months. A sow, therefore, may give birth to twenty young in the same period of time required by the mare for the development and birth of one offspring.

86. Duration of the reproductive period. — The fertility of an individual or a breed is largely determined by the duration of the reproductive period. The length of this period is a matter of special importance in the larger animals which produce but one young at a birth. The breeding age is the time from the arrival of puberty until the cessation of the breeding function on account of old age. The beginning of the reproductive period has already been discussed under puberty. The average age at which the domestic animals cease to breed is difficult to determine because of the scarcity of records on this point. The persistence of the breeding powers of the domestic animal is influenced by the conditions under which they are reared. With favorable conditions, the mare has produced young regularly until twenty-five years old. Ewes have continued to produce offspring until nineteen years of age.

87. Confinement and fertility. — The close confinement of animals seriously interferes with their fertility. Wild animals like the elephant, tiger, lion, squirrel and monkey in captivity are often sterile. Darwin states that the same animals breed more readily in traveling shows than in zoölogical gardens. Among the domestic animals, confinement and lack of exercise are frequent causes of a low degree of fertility. It is generally a mistake to confine ewes or sows in a small lot during the winter months, under conditions which make it impossible for them to secure adequate exercise. The same may be said of males, including stallions, bulls, boars and rams. The comparative infertility of a stallion may often be traced directly to the fact that he is kept stabled in a small dark stall with no regular exercise. Mares both before and after breeding should be given regular exercise. It is better that they should work at some slow regular task every day.

In an investigation on the relation of exercise to the development of the internal organs by Kulbs and Berberrich,[1] it was found that in the case of dogs and swine the size of the muscles and the weight of the heart and liver were increased by exercise. It was also observed that the color of the bone marrow was deepened.

[1] " Jahrbuch Wiss. und Prakt. Tierzucht," 6 (1911), p. 232.

Darwin[1] reports that "It has been found in France that with fowls allowed considerable freedom, only twenty per cent of the eggs failed; when allowed less freedom, forty per cent failed; and in close confinement, sixty out of the hundred were not hatched."

88. The fertility of domesticated animals. — From what has already been said regarding confinement and its deleterious effect on fertility, it might be concluded that domestication is unfavorable to high fertility in animals. This is far from the truth. Domestication furnishes or ought to furnish the most favorable conditions for high fertility in animals. A regular and abundant supply of nutritious food, shelter from a rigorous climate, and the opportunity for selection by man all contribute to the development of a relatively high degree of fertility in animals. All races of domestic animals are more fertile than their wild prototypes. Darwin has pointed out that the tame rabbit gives birth to four to eleven in a litter and breeds six or seven times a year, while the wild rabbit only produces five or six young at one time and breeds four times yearly. The domestic fowl may produce as many as 200 or more eggs in one year, while the female of the wild progenitor of the domestic hen lays only six to ten eggs in a year. The same remarkable difference exists between the domestic and wild duck. The Indian Runner duck under domestication will produce 250 eggs a year, and the wild duck only five to ten eggs in one year. At the Government Experiment Station of New South Wales in Australia, six Indian Runner ducks laid 1601 eggs in one year, an average of 267 eggs each.

[1] Darwin, "Animals and Plants under Domestication," vol. II.

89. Age and fertility. — The fertility of animals is influenced by age. Young animals that have not yet reached maturity are generally less fecund than mature individuals. There is also some evidence to show that old mature animals are more prolific than younger mature animals. It is highly probable that skillful feeding and management may result in a significant increase as age advances.

90. Relation of age to fertility in swine. — At the Missouri Experiment Station, the author has had under investigation the relation of the age of animals to their breeding powers. The animals used in this experiment have been swine, and the general plan has been to divide the sows into three groups according to age. Group I is composed of very young sows from four to five months of age; group II of half mature sows about eighteen months of age; and group III of mature sows from twenty-four to thirty months old. The young immature sows of group I have been bred at the first appearance of puberty, which in well-fed sows is from four to five months old. The female offspring of the immature sows are again bred at the first appearance of heat, and this will be continued indefinitely. A number of interesting phenomena have been observed, but only those results which throw light on the relation of age to the fertility of the mother are recorded here.

91. Influence of age of sow on size of litter. — In the Missouri experiment described above, there have been to date (1913) twenty-six litters from immature sows. The average number of pigs to the litter has been four and eight-tenths. From the half mature sows, eight litters have resulted in an average of six and three-tenths to a litter. The average number of pigs in six litters from the fully mature sows has been six and five-tenths.

The sows used in this experiment were all of the same breeding, and received the same care, and the food, shelter and all other conditions have been similar. The differences observed then must be due to the factor of age. It will be observed that there is a marked difference between the size of the litters of the immature sows and the older ones. The litters of the older sows are materially larger than from the immature sows.

The highest fertility in swine is not reached until the mother is at or near maturity. It may not always be profitable for the swine-breeder to delay breeding sows until full maturity, but it is apparent that when the breeding herd is composed of older sows, a smaller number need be maintained for the production of the pigs needed in a given system of farm management.

Interesting statistics have been compiled by George M. Rommel[1] from the records of the American Poland-China Record Association, which have an important bearing on this point. These statistics include the breeding records of 6145 sows recorded in 1902. There were examined the breeding records of 2010 one-year old sows. The litters of 1520 one-year-old sows, or seventy-five per cent, ranged from five to eight pigs.

The average litter of 2047 two-year-old sows numbered seven and five-tenths. The number in each litter of 1483 sows, or seventy-two per cent, ranged from six to nine pigs.

The average number of pigs in the litters of 1157 three-year-old sows was seven and nine-tenths. The average litter recorded for 606 four-year-old sows was

[1] American Breeders' Association, Report 19.

eight and three-tenths. The records for 325 five-year-old sows give the average size of litter as eight and seven-tenths. The total number of sows examined, ranging in age from one to five years, was 6145 and the average size of litter for the whole number was seven and four-tenths.

The number of animals included in these investigations and the unquestioned accuracy of the records make these figures valuable. It is clear that the sows increase in fertility from one to five years. From the standpoint of the practical breeder, it would seem that sows from two to four years of age will be most profitable from the standpoint of prolificacy.

92. Relation of age to fertility in sheep. — The greater fertility of the older females has been noted in ewes by the Wisconsin Experiment Station in Bulletin No. 95. Observations made at that station on the percentage of increase from ewes of different ages show that two-year-old ewes gave an annual increase of 158 per cent, three-

THE EFFECT OF THE AGE OF EWES ON PER CENT OF INCREASE AND SEX OF LAMBS

WISCONSIN EXPERIMENT STATION, BULLETIN 95

AGE OF EWE BEARING	TWO YEARS		THREE YEARS		FOUR YEARS		FIVE YEARS		SIX YEARS		SEVEN YEARS	
	No.	Per cent	No.	Per cent	No.	Per cent	No.	Per cent	No.	Per cent	No.	Per cent
Single Lambs	62	44.6	30	31.9	21	30.4	14	28.5	9	33.3	6	60.0
Pairs Twins	72	52.5	58	61.7	42	60.8	32	65.3	15	55.5	3	30.0
Sets Triplets	4	2.9	6	6.4	6	8.8	3	6.2	3	11.2	1	10.0
Rams	96	49.0	75	51.0	71	57.2	45	51.7	24	53.3	10	66.7
Ewes	100	51.0	72	44.0	53	42.8	42	48.3	21	46.7	5	33.3
Per cent increase		158.0		174.0		178.0		177.0		178.0		150.0

year-old ewes an increase of 174 per cent, and four- to six-year-old ewes an increase of 178 per cent. After the age of six years, there was a distinct falling off in the percentage of increase.

The foregoing table (page 93) is reprinted entire from the Wisconsin bulletin and is interesting as showing the distribution of twin and triplet births among the different ages and the sex of lambs.

93. Influence of age of ram on fertility of ewes. — The number of young at a birth is generally believed to be determined by the number of ova which are ripened by the female during any one period of heat. The vigor or age of the male used is not generally regarded as having any influence in determining the number born at one time. The Wisconsin Station [1] found that during a period of six years, the flock of ewes served by a yearling ram produced 150 per cent of lambs. The same flock during a similar period of six years was served for three of these years by two- and three-year-old rams. The average percentage of lambs born from the older rams was 180 per cent. The author in discussing these results remarks: "These data are quite at a variance with the opinion commonly held by sheepmen, generally to the effect that a well grown, vigorous yearling ram is at his best as a sire. It is also contrary to the belief held by many that the vigor of the sire has no apparent influence on the percentage of increase."

94. The effect of the age of poultry parents on the offspring. — The general conclusion that fully mature animals are more fertile seems to be substantiated by the poultry-breeding experiment conducted by Atwood.[2]

[1] Wisconsin Experiment Station, Bulletin 95.
[2] Atwood, West Virginia Experiment Station, Bulletin No. 134.

THE EFFECT OF AGE OF THE PARENTS UPON THE VIGOR OF CHICKENS[1]
WEST VIRGINIA AGRICULTURAL EXPERIMENT STATION

	Test 1			Test 2			Test 3			Av. 1	Av. 2	Av. 3
	Pen 1	Pen 2	Pen 3	Pen 1	Pen 2	Pen 3	Pen 1	Pen 2	Pen 3			
Number of eggs put in incubator	55	79	74	114	92	88	71	55	57	80	75.3	73
Weight of eggs per 100 lbs.	11.5	12.18	14.29	11.71	12.22	13.16	12.21	12.04	13.07	11.8	12.1	13.5
Tested out infertile	8	9	5	19	15	9	18	7	8	15	10.3	7.3
Unhatched or cracked	2	10	9	—	4	5	—	—	—	—	—	—
Number of chickens	45	60	60	76	62	64	51	46	39	57.3	56	54.3
Per cent hatched (except cracked)	81.8	75.9	81.1	66.7	70.5	77.1	71.8	83.6	68.4	73.4	76.6	75.4
Weight of chickens per 100 lbs. when taken from incubator	7.33	7.54	8.47	7.64	7.75	8.50	7.64	77.1	8.0	7.57	7.66	8.65
Early deaths	21	8	6	12	7	3	5	1	3	12.7	5.3	4

Breed.— Single Comb White Leghorns.
Pen 1.— Pullets.
Pen 2.— Hens Two Years Old.
Pen 3.— Hens Three Years Old.

[1] Atwood, American Breeders' Association, Vol. 5.

In this investigation, Single Comb White Leghorns were employed and a comparison of hens, pullets, two-year and three-year-old hens was made. The important results may be seen at a glance from the table (page 95).

The records of this test show that the eggs laid by old hens are heavier than those laid by pullets, that the number of chicks hatched was ten per cent greater, that the initial weight at hatching time, and for several weeks thereafter, was greater from the older hens, and finally that the percentage of chicks dying from pullet's eggs was three times greater than from the mature hens.

In marked contrast to the above results are those published by the Maine Experiment Station in Bulletin 168. The author (Pearl) of this publication states, "The present statistics do not show any marked superiority of hens over pullets in respect to breeding performance, so far as either fertility or hatching quality of eggs are concerned."

95. Age and fecundity. — Duncan[1] distinguishes between the ability to bear children, which he calls fecundity, from actual productiveness or the number of births, which is designated as fertility. From Duncan's investigations it is possible to formulate a general law which represents a true statement of the relation of age to fecundity. This general law has been stated by Marshall[2] as follows: "The fecundity of the average individual woman may be described, therefore, as forming a wave, which, starting from sterility, rises somewhat rapidly to its highest point and then gradually falls again to sterility." The results discussed earlier in this chapter clearly indicate that in

---

[1] Duncan, "Fecundity, Fertility, Sterility and Allied Topics," Edinburgh, 1866.
[2] Marshall, "The Physiology of Reproduction," p. 590.

the case of swine this law is undoubtedly a true statement of what actually happens. This law not only applies to mammals but is also found in poultry. Geyelin [1] has attempted to formulate an average of fertility in poultry in relation to age in the following table:

First Year after hatching	15 to 20 eggs
Second Year after hatching	100 to 120 eggs
Third Year after hatching	120 to 135 eggs
Fourth Year after hatching	100 to 115 eggs
Fifth Year after hatching	60 to 80 eggs
Sixth Year after hatching	50 to 60 eggs
Seventh Year after hatching	35 to 40 eggs
Eighth Year after hatching	15 to 20 eggs
Ninth Year after hatching	1 to 10 eggs

These estimates must be regarded as far below the performance of well-selected flocks maintained under good conditions of food and shelter. The age at which pullets begin laying varies greatly, depending upon their development. At the Ohio Experiment Station, a White Leghorn pullet began laying at four months and fifteen days old. At the Missouri Experiment Station Kempster [2] reports a White Leghorn pullet beginning to lay at four months and nineteen days of age. The number of eggs laid by old hens may also greatly exceed the figures given by Geyelin. At the Maine Experiment Station a hen laid 111 eggs during her ninth year.[3]

A remarkable case of fecundity in sheep is noted by Pearl.[4] A ewe owned by Barrett for nineteen years gave birth to thirty-six lambs which were distributed during the breeding life of the ewe as follows (page 98):

[1] Geyelin, quoted by Marshall, *loc. cit.*, p. 590.
[2] Unpublished records, Missouri Experiment Station.
[3] Maine Experiment Station, Bulletin 266.
[4] Pearl, *Science*, vol. 37, p. 227.

		LAMBS
April,	1806	1
	1807	1
	1808	2
April 3,	1809	3
March 29,	1810	3

Making 6 lambs in 11 months and 26 days

	LAMBS
1811	3
1812	3
1813	3
1814	3
1815	2
1816	2
1817	2
1818	2
1819	2
1820	2
1821	1
1822	1
1823	0
1824	0
Total	36

Pearl has further called attention to the fact that " the median point in the breeding career of this ewe was 8.17 years. That is, she produced one-half of her offspring before and one-half after it." The age of maximum fecundity in this ewe was 7.34 years.

96. Nutrition and fertility. — All the physiological activities of an animal are influenced by nutrition. This is particularly the case with the reproductive functions. The fertility of an animal is influenced by the kind and the amount of food consumed. Certain kinds of food have long been believed to affect injuriously the breeding functions of animals. Sugar fed to domestic animals in considerable amounts has apparently had an unfavorable influence on fertility.[1] Greatly increased activity

[1] Tanner, *Journal of Royal Agricultural Society*, 1865, p. 267.

of the generative organs is characteristic of the spring season. At this season most animals, domestic and wild, are periodically in heat. This greater activity is undoubtedly due to the abundant supply of nutritious and succulent grass. That this kind of food does materially influence the fertility of animals is recognized by the shepherds of England in the practice of flushing ewes. This practice consists in turning the ewe flock on rich succulent pastures about two weeks before turning in the ram. The flock owners believe that this increases the number of lambs and brings the ewes more uniformly in heat. The Beinn Bhreagh [1] flock of sheep in Nova Scotia belonging to Dr. Bell when fed generously before and during the mating season produced a larger number of twins. The older ewes also produced a larger percentage of twins.

97. Excessive food supply and nutrition. — An oversupply of nutritious food which causes the animal to become abnormally fat is often the cause of sterility among the domestic animals. In nature it is rare for an animal to remain continuously in an excessively fat condition. At certain seasons of abundant food supply the wild animal may become fat, but such periods of plethora are invariably followed by a scarcity of food, and such food must often be gathered by exhaustive exercise. Such variations, if not extreme, may be particularly favorable for the functioning of the reproductive system. Certain it is that the most skillful stockmen have long recognized the fact that the female reproductive functions are most active when the individual is actually gaining in condition. An animal that is maintained in a uniform condition of excessive fatness is not

---

[1] Bell, *Journal of Heredity*, vol. V, p. 47.

in the best condition for the successful exercise of the breeding function. A rapidly improving condition due to nutritious food supplied in generous quantities is distinctly favorable, provided the animal is not already too fat.

E. Davenport has held that, "excessive food supply leads to infertility among both plants and animals." This is true of long-continued and excessive feeding, but a rapidly improving condition of the animal in thin condition is distinctly favorable to the highest fertility. It is a mistaken idea that starvation or a very limited diet is a favorable environment for the successful activity of the generative system. Such treatment is only favorable in the case of over-fat animals and is oftener the first and a very essential step in securing offspring from animals that through a long period of overfeeding become temporarily barren.

98. Other factors affecting fertility. — A sudden change of conditions surrounding the animal, such for example as the exportation of an animal from Europe to the United States, will often temporarily interfere with the normal activities of the reproductive system and the animal may be barren for a time. When the animal has become thoroughly accustomed to the changed conditions, its breeding powers return and thereafter may function normally. Changed conditions may also result in increased fertility. As shown elsewhere, breeding animals in thin condition and existing upon a sparse ration or upon a dry dietary, become markedly more fertile when changed to richer pastures.

Some individuals are infertile when mated with certain other individuals, but may be fully fertile with others. A mare may be sterile when bred to a stallion, but fertile when mated with a jack. Such a physiological aversion is not easy to explain, but is nevertheless so frequent as to be a well-recognized fact among breeders. Prolonged lactation must be regarded as unfavorable to fecundity in some species. This is especially true in the case of swine, where early weaning of the litter will certainly encourage an earlier return of the heat period and thus make possible a larger number of litters during the natural breeding life of the mother.

99. Relation of number of mammæ in swine to fertility. — An interesting study of the mammæ in swine was reported by Wentworth.[1] From these researches there is little evidence in favor of the popularly accepted opinion that there is a relation between the fertility of swine and the number of mammæ. The normal type of mammary pattern in swine consists of regularly placed pairs on the ventral side of the body. The first pair lie immediately behind the juncture of the ribs and sternum. The greatest variation occurs in the second pair. The last pair are closer together and thus nearer the median line in an inguinal position. Variations occur in the number of pairs and also in the suppression of one nipple of a pair. These variations are often inherited. The normal number of mammæ in the Tamworth and Berkshire breeds is 13, 14 and 15, in the Duroc-Jersey breed 10, 11 and 12. The tendency to vary is greater when the number of pairs exceeds five.

100. Twins. — The normal number of young in several of the larger breeds of the domestic animals and in man is one. The production of a larger number at a single birth is exceptional. It happens, however, that twins are frequently born, while triplets and even four and

[1] Wentworth, *Amer. Naturalist*, vol. 47, p. 257.

five at a birth have been reported. When twins are born they are either of identical sex or one a male and the other a female. In some cases the twins are very much alike in all other characters as well as sex. Such twins were called by Galton identical twins. It is also true that twins are often born which have no greater resemblance to one another than ordinary brothers and sisters. Such twins undoubtedly develop from separate eggs and are known as ordinary or fraternal twins. They do not necessarily resemble one another more closely than brothers and sisters of the same family except that they are of identical age and for this reason might be expected to have a closer resemblance than brothers or sisters of widely different ages. Fraternal twins may be of different sex. Identical twins are believed to come from one egg after fertilization. They are always of the same sex and very much alike in external and internal character and in mental and moral tendencies.

101. Characters correlated with fertility. — It is in the highest degree desirable that the breeder should be able to distinguish those qualities, external and internal, which are in any way correlated either with fertility or sterility. Unfortunately we cannot now speak with assurance on all the supposed evidences of fertility in animals, but some characters are undoubtedly closely correlated with fertility and we may through them learn to judge of the probable existence or nonexistence of this most desirable trait. Manifestly, characters closely correlated with fertility will finally persist and become dominant. It is equally evident that those characters of the animal body which are correlated with infertility will ultimately disappear. The skillful judge of breeding animals recognizes some such correlation in the selection of both male and female individuals. The breeder emphasizes the existence of those general qualities which give to the male a distinctly masculine appearance and to the female a clearly recognizable character of femininity. The bull possessing a markedly masculine aspect is held to be a " good breeder." Whether it is meant by this that such a bull is prepotent in fixing his own characteristics upon his offspring or, what is more probable, that a bull of this character is more than ordinarily efficient in the development of sperm-cells, it is still true that the masculine type may be regarded as in some degree at least correlated with fertility. Supernumerary mammæ have been found in many cases associated with exceptional fertility. In describing the dam of triplet calves, Pearl[1] remarks: " It is of interest to note that this cow has two very small posterior mammæ. It is of course impossible to say whether this occurrence of supernumerary mammæ is directly connected with the high degree of fecundity exhibited by this cow, but this may fairly be regarded as probably the case because of the fact that these two things are known to be associated in other forms." The sheep breeding experiments at Beinn Bhreagh[2] by Alexander Graham Bell have suggested a possible correlation between extra nipples and unusual fecundity. Stature and fertility have been found by Pearson[3] to be somewhat closely correlated among women. The taller women are on the average more fertile. If this is generally true, the stature of women is likely to increase at least until it has reached a point which satisfies the correlation existing. Among swine-breeders it is generally

[1] *Loc. cit.*
[2] Bell, *Science*, N. S., vol. 36, pp. 378–384.
[3] Pearson, "Grammar of Science," pp. 441–445.

believed that sows with rather long bodies are more fertile than shorter, more compact individuals. It seems to be true also that females having a somewhat loose and open conformation are generally more certain breeders.

The milking function of animals is in a measure correlated with the quality of fecundity. Breeds of animals and individuals which have the milking function well developed are more fecund than those in which the development of this quality has been neglected. Tanner,[1] in his interesting discussion on "The Reproductive Powers of Animals," says: "The formation of milk is intimately correlated with the reproductive powers. The secretion of milk is dependent upon the activity of the mammary glands and these are either under the direct influence of the breeding organs or else they sympathize very closely with them. Those animals which breed with the least difficulty yield the best supplies of milk and produce the most healthy and vigorous offspring." He also adds that, "Since a short supply of milk is indicative of and associated with enfeebled breeding powers every care should be taken to obviate this defect."

It must be admitted that our knowledge on the subject of characters correlated with fertility is as yet fragmentary and indefinite. The importance of this quality in practical breeding should make this a fruitful field for further investigation.

102. In-breeding and fertility. — Continuous in-breeding among domestic animals has in many instances been followed by low fecundity or absolute sterility. It is generally believed by practical breeders that of all the ill effects supposed to result from in-breeding, lessened

[1] Tanner, "The Reproductive Powers of Animals," *Journal of Royal Agricultural Society*, 1865, p. 270.

fertility is the one most likely to follow. Whether this loss of fecundity in animals of consanguineous breeding is to be attributed to in-breeding *per se* or whether it is due to the rapid fixing of a tendency to sterility already existing in the family, it is nevertheless true that there exists a certain amount of probability that continuous close-breeding will ultimately affect injuriously the fertility of animals. This question is discussed at some length under " In-breeding," Chapter XI.

103. Cross-breeding and fertility. — It naturally follows that if in-breeding is unfavorable to full fecundity, cross-breeding must tend to develop this desirable quality. Here again it is not easy to trace the increased fertility which follows the mating of animals of diverse characters to the sole act of crossing. Many of the cases of increased fecundity due to crossing may be explained on the basis of introducing the new quality of high fertility. If fertility is a dominant character transmitted in accordance with the Mendelian principle of dominance, it is easy to understand why the cross-bred animal may exhibit, as it often does, a greater degree of fecundity than can be accounted for on the assumption of blended inheritance of this quality from both parents. What actually happens is that one of the parents possesses the quality of fertility in high degree and this becomes dominant in the offspring. A certain proportion of the offspring, therefore, receive in the constitution of the germ-plasm all the high fertility which is an inherent part of the germ substance of the one parent. It should also follow that a certain proportion of the offspring inherit unchanged the tendency to low fecundity characteristic of the other parent. It must be admitted that evidences of the latter are still lacking, but it should be possible by experiment to determine this point.

104. Unusual fertility. — Each species and most varieties or breeds of animals have a fairly uniform and normal rate of increase. Thus the normal number of young at a birth in cattle and horses is one. It is also probably true that among sheep one is the normal number of young at each birth. But many breeds of sheep have been so changed by domestication that twins are frequent and triplets are not rare. The quality of fertility is undoubtedly transmitted by heredity. It is therefore possible to increase the normal fertility of the domestic sheep by selection. Among cattle the production of twins has not been regarded as a particularly desirable quality, and hence no attempt has been made to increase the normal birth number of the bovine species. It is not difficult, however, to conceive that it would be comparatively easy to develop a breed or variety of cattle which would produce twins. Cases of unusual fertility among all classes of the domestic animals are frequent. These are of enough importance from a practical point of view and of sufficient biological significance to be given a place in a discussion on fertility.

105. Unusual fertility among horses. — Exceptional fertility among horses is generally to be found in connection with longevity and active and regular functioning of the breeding powers rather than in unusual numbers of young at a birth. A mare twenty-five years old, owned by R. O'Heren of Illinois, in 1904 was suckling her twentieth colt. She was one-half Thoroughbred and one-half Clydesdale and was still strong and active.[1] The Thoroughbred mare, Fanny Cook, dam of Daniel Lambert, produced fifteen foals and dropped twins at twenty-two

---

[1] Reported in a letter to the author by R. H. Dunn, Illiopolis, Ill.

years of age. A Clydesdale mare belonging to G. W. Henry of Burlington, Iowa, was the mother of nineteen foals and was supposed to be in foal again.[1] Pocahontas, a running mare, was the mother of fifteen foals and dropped her last at the time she was twenty-five years of age.

106. Unusual fertility among cattle. — Some remarkable cases of fecundity among cattle have been recorded. In most cases the ability to ripen a number of eggs during one period of heat seems to be inherent. A cow that has produced twins or triplets is very apt to do so again. This tendency to multiple gestation in cattle is well illustrated by a family of cattle on a New Hampshire farm reported by Wentworth.[2] "The foundress of the family was a grade Holstein cow, herself a twin, about seven years old. She has been on the farm ever since she was dropped and has given birth to seven calves. Her first service was to a Guernsey bull and resulted in a pair of yellow and white heifers, one of which is now in the herd. Her second mating to a red Shorthorn bull resulted in a single black and white bull calf that was vealed. An Ayrshire bull sired her third calves, twin black and white bulls, but neither of these was good enough to raise. Her fourth service was to a Holstein bull and from it she produced twin black and white heifers that promise well as milkers."

"The yellow and white twin first produced by the old cow is now four years old and has twice borne twins. To an Ayrshire bull she produced a pair of yellow and white bull calves that early went to the butcher and to a

---

[1] Sanders, "Horse Breeding," p. 179.

[2] Wentworth, E. N., "Twins in Three Generations," *Breeder's Gazette*, vol. 62, p. 133.

Holstein bull she gave birth to twin black and white heifers last December."

Pearl[1] has described an interesting case of triplet calves from a grade Guernsey cow seven years old. The sire was a young grade Hereford bull which had not shown any unusual tendency to sire twins or triplets. Two of the calves were heifers and had the typical white face of the Hereford breed. The third calf, a bull, was a typical Guernsey and in color and coat markings resembled somewhat closely his dam. (Plate IV.)

In reference to the breeding record of these triplet calves, Pearl says: "The writer asked Mr. Walter, the owner of the calves here described, to pay particular attention to the sexual behavior of these triplets. This was done. As has already been implied in what has gone before, the male individual of the triplets was entirely functional sexually. He was used in service locally; got good calves; and apparently got as high a proportion of calves as would be expected from a bull of his age. In regard to the sexual history of the female individuals of the triplets, Mr. Walter has the following to say in a letter dated April 11, 1910. After noting the fact that these two supposed heifers had been killed and sold in the village market he says: 'Neither of them had ever been in heat.' In earlier letters Mr. Walter on several occasions said that these calves never showed the slightest signs of being in heat. From the account given by the butcher who killed these animals it appears probable that in both individuals the conditions were such as have been described for many free-martins. Neither uterus or tubes were recognized, but the vagina apparently

[1] Pearl, "Triplet Calves," Bulletin 204, Maine Agricultural Experiment Station.

ended at its anterior end as a blind sack. Although detailed anatomical data are lacking, there can be little doubt, I believe, because of both physiological fact and absence of œstrus and the lack (?) or minute, infantile condition of uterus and Fallopian tubes, that these two supposed female individuals were really freemartins."

The cow possessed two supernumerary mammæ just behind the posterior pair. The occurrence of supernumerary mammæ has before been observed to accompany the tendency to multiple births.

A seven-year-old Shorthorn cow dropped seven dead calves at one birth. They were sired by a Holstein bull.[1] A Shorthorn cow three years old, on post mortem, was found to be carrying six perfectly developed calves in her uterus.[2] Another Shorthorn cow gave birth to four calves, three of which were weak and undeveloped.[3] A cross-bred cow gave birth to seven calves within a period of twelve months. All these calves were born alive.[4] A cow dropped three pairs of twins in succession during a period of two years.[5] A grade Guernsey cow on a farm in Washington County, Pennsylvania, gave birth to triplets. This cow was ten years old and had produced fifteen calves at eight births. A cow twenty-two years old is reported as having had twenty calves and was again pregnant.[6] A remarkable case of continued high fertility in a cow is quoted by Pearl from McGillwray's "Manual of Veterinary Science and Practice." The cow

---

[1] *Country Gentleman*, 1895, p. 595.
[2] *Ibid.*, 1880, p. 313.
[3] *Ibid.*, 1891, p. 339.
[4] *Ibid.*, 1893, p. 231.
[5] *Breeder's Gazette*, 1898, p. 7.
[6] *Rural New-Yorker*, 1906, December.

described was of "the black polled breed" and her record of births follows:

Year	Number of Calves at Birth	
1842	1	This the cow's first calf
1843	3	All lived to adult age
1843	4	One died. (Seven calves in one year)
1844	2	Lived to maturity
1845	3	Lived to maturity
1846	6	All died prematurely
1847	2	Came to maturity
1848	4	
Total	25	Mean number per birth — 3.125

107. Unusual fertility among sheep. — Sheep normally produce a larger proportion of twins than cattle or horses. This may be due in a measure to the fact that the sheep has a much shorter period of gestation. It is true in general that those mammals having the shortest periods of gestation are most prolific. Some remarkable cases of great fertility among sheep are matters of record. A three-year-old grade Cotswold ewe gave birth to five fully developed lambs. Two died at birth, the others in a few hours.[1] A Horned Dorset ewe four years old dropped five lambs at two births within a ten months' period.[2] A Radnor ewe dropped six lambs at one birth, of which five lived and thrived.[3] An Oxford-down ewe gave birth to four strong, vigorous lambs which grew rapidly and weighed one hundred and sixty-two pounds at eight weeks old.[4] A prolific ewe at one birth dropped five lambs, all of which were perfectly developed and grew rapidly.[5] A Leicester ewe gave birth to six strong, healthy

---

[1] *Country Gentleman*, 1893, p. 171.
[2] *Breeder's Gazette*, 1894, p. 327.
[3] *Country Gentleman*, 1892, p. 331.
[4] *Breeder's Gazette*, 1893, p. 388.
[5] *Country Gentleman*, 1878, p. 329.

lambs, four of which the mother nursed successfully. The earless Shanghai breed of sheep exhibited in the London Zoölogical Gardens in 1857 seem to have inherited a remarkable fecundity. Bartlett [1] has described this variety as breeding twice each year and often producing four or five at a birth. In the spring of 1857 three ewes of this breed gave birth to thirteen lambs.

108. Unusual fertility among swine. — There is more difficulty in determining the normal number of young at a birth among swine than among other domestic animals. There is considerable variation among individuals belonging to the same breed and between different breeds. Some particular cases of high fertility are described below:

A three-year-old Chester White sow [2] farrowed ninety-six pigs in six litters. There were fourteen in each of the first three litters and eighteen in each of those last farrowed. This tendency among highly fecund individuals to give birth to larger and larger numbers has been observed in cattle, sheep and swine.

A Poland China sow [3] produced thirty-four living pigs in three litters during a single twelve months' period.

A sow [4] gave birth to twenty-one pigs in a litter. Previous to this she had farrowed two litters of fifteen and seventeen pigs each. A sow of uncertain breed [5] dropped twenty-three pigs in one litter. All but two of these were born alive. The same sow gave birth to eighty-five pigs in five litters. Ray L. Zimmerman of Amazonia, Andrew County, Missouri, reports to the author that a

---

[1] Bartlett, *Proc. Zoöl. Soc.*, London, 1857, p. 105.
[2] *Breeder's Gazette*, 1897, p. 368.
[3] *Ibid.*, 1894, p. 308.
[4] *Country Gentleman*, 1887, p. 281.
[5] *Ibid.*, 1894, p. 915.

Poland China sow owned by W. Minner of that county farrowed twenty-five pigs in one litter.

109. Unusual fertility among poultry. — An instance of remarkable ability in egg-laying is given in the Experiment Station Record, vol. 28, p. 270. In this volume is described a Single Comb Brown Leghorn hen which laid 257 eggs in twelve months. This hen weighed only three and two-tenths pounds. The average weight of the eggs was one and eight-tenths ounces. At the Delaware Experiment Station a White Leghorn hen, Lady Eglantine, laid 314 eggs in one year.

CHAPTER VI

STERILITY

It goes without saying that the first essential quality in a breeding animal is the ability to produce young. The more highly developed the animal is in those special characters which have been fixed by selection, the more important becomes the mere ability of an individual to give birth to offspring. In an animal reared primarily for commercial purposes, like the hog or the beef type, the barren individual is not so serious a loss, as it may still have a value for meat.

It is well known that many individuals among the domestic animals are sterile. Such sterility is found among animals reared under the best conditions, as well as among those subjected to less skillful husbandry. Barrenness occurs in individuals which are a part of herds or flocks in which all animals are surrounded by identically the same conditions. It must be true, therefore, that some animals possess a tendency to barrenness in a more marked degree than others. This tendency may possibly be inherited. Barrenness may be only temporary or it may be a permanent condition. When it is a temporary condition, it can often be alleviated by knowing the conditions which are favorable to fertility, and particularly those conditions which are known to act unfavorably upon the breeding functions.

110. The causes of sterility. — The various causes[1] of sterility may be classified as anatomical, physiological, pathological or psychological. Sterility in the male may be due to an inability to perform the sexual act, which condition is known as impotence, or it may be due to an inability properly to develop spermatozoa.

111. Causes of sterility in the male. — The male may be sterile as a result of undeveloped testicles as in some ridglings, where the testicles are retained in the abdomen. The ridgling is not always permanently sterile and may be fully fertile, but the failure of the testicles to descend normally from the abdominal cavity into the scrotum is to be regarded with suspicion by the breeder. The only way to determine whether a particular ridgling is fertile is by actual trial. There is no medicinal or surgical treatment which can make a barren ridgling fertile.

Bulls, boars and stallions which are fed upon a generous ration of highly nutritious food and are not given regular exercise tend to become over-fat, and such a condition often leads to fatty degeneration of the testicles and consequent sterility. This condition is recognized by all successful breeders. Breeding males that have proven themselves of great merit as sires and are not intended for exhibition are generally and wisely maintained on a moderate allowance of nutritious food, being careful to limit the amount and provide some means for exercise, thus avoiding the almost inevitable fatty degeneration of the reproductive tissues which follows long-continued high feeding combined with little exercise. An example of the intimate relation of these factors to fertility is to be observed in the breeding practices of many modern

[1] *See* "Diseases of the Horse," U. S. Department of Agriculture, 1907, pp. 151–154.

owners of draft stallions as compared with the methods of early stallioners in the newer agricultural sections of the United States. It was formerly the custom to drive or ride the stallion from farm to farm, thus often covering a territory of 100 to 200 square miles. Stallions so handled were notoriously sure foal-getters and not infrequently were successful in getting from eighty-five to ninety-five per cent of the mares in foal. The modern plan of keeping the draft stallion in high condition and standing him for service at one barn, thus requiring all mares to come to him, has undoubtedly reduced the fertility of draft stallions. It is no unusual event for a draft stallion so managed to get only sixty per cent of mares in foal, while seventy-five per cent of the mares in foal is regarded by some stallioners as a fair average for stallions handled in this manner. The fatty degeneration of the vasa deferentia or excretory ducts of the testicles is also a not infrequent cause of sterility in very fat animals.

Any injury or deformity of the penis which renders the act of copulation painful or impossible is to be included in the category of anatomical causes of sterility. In this class may be included inflammation or ulceration of the mucous membrane enclosing the penis, paralysis of that organ, and the presence of tumors on the penis itself or its appendages. Muscular and bone diseases which in any way interfere with the exercise of the breeding function are causes of barrenness. Spavin or ring bone may cause the stallion such inconvenience and distress in mounting as to prevent copulation and thus indirectly be a cause of barrenness. Similarly, diseases of the muscles of the back and loins may be responsible for sterility in certain individuals. Diseases of the brain and spinal cord, particularly those which control the act of coition,[1] diabetes and albuminuria are to be regarded as causes of sterility.

The potency of the semen of the male may be so weakened by too frequent services by the stallion as to result in sterility. The number and frequency of services which can be required of a stallion and still retain the full vitality of the sperm-cells varies greatly with different individuals. An interesting contribution to this subject is reported by Lewis:[2] A draft stallion was permitted one cover daily for nine consecutive days. Samples of the semen were taken to the laboratory in a warm sterilized receptacle. The number of sperm-cells to the cubic millimeter in the semen from the first service was 131,750. The number of sperm-cells decreased daily with considerable uniformity until the ninth service, when the number suddenly diminished from 51,480 to 5840. The vitality of the spermatozoa was determined by maintaining the fluid at a uniform temperature and determining the number of viable sperm-cells at the end of a given period. Thus when the semen was kept at a temperature of 31 to 35° C. it was found that five per cent of the cells from the first service were alive after nine and five-tenths hours. From the third service no cells were alive after six hours. From the sixth service no cells were alive after four hours, and from the eighth service no cells were alive after three hours. In a second test a grade stallion was bred eleven times on consecutive days. The author summarizing this test concludes that:[3] "Approximately the vitality of the cells decreased one-

---

[1] Comer, "Diseases of the Male Generative Organs," 1907.
[2] Lewis, Oklahoma Experiment Station, Bul. 96, 1911.
[3] *See* Marshall, "The Physiology of Reproduction."

half and the number to one-fifth in the last or eleventh service of the series as compared with the condition of the semen at the first service." Other conditions which are to be regarded as sources of barrenness are incomplete erections, premature ejaculations, fear and repugnance. The inability of the male to produce fertile semen may be a congenital condition or it may be acquired through some of the causes mentioned in this chapter.

112. Sterility in the female. — The failure of the female to produce offspring is due to a variety of causes. Some of these are inherent and cannot be successfully treated. Such animals are permanently barren and of course useless for breeding purposes. There are other causes of infertility which are the result of purely temporary circumstances, and these may often yield to skillful treatment by man and valuable animals thus become regular breeders. Some of the more important causes of sterility which are temporary and which may yield to treatment are mentioned below.

113. Closure of the cervix. — Mares and cows often fail to become pregnant as a result of the constriction of the muscles forming the neck of the womb. This spasmodic closure of the cervix prevents the passage of the semen from the vagina into the uterus, and the fertilization of the egg is thus prevented. This condition is more frequent in young females that have never been pregnant, but is not uncommon among animals that have previously given birth to offspring. This condition may generally be successfully treated by a simple operation known to the stallioners as opening. The treatment for this condition is admirably and clearly described by Law[1] as follows: "Spasmodic closure of the neck of the womb

[1] "Diseases of the Horse," U. S. Department of Agriculture, 1907.

is common and is easily remedied in the mare by dilatation with the fingers. The hand, smeared with belladonna ointment and with the fingers drawn into the form of a cone, is introduced through the vagina until the projecting, rounded neck of the womb is felt at its anterior end. This is opened by the careful insertion of one finger at a time, until the fingers have been passed through the constricted neck into the open cavity of the womb. The introduction is made with a gentle rotary motion, and all precipitate violence is avoided, as abrasion, laceration or other cause of irritation is likely to interfere with the retention of the semen and with impregnation. If the neck of the womb is rigid and unyielding from the induration which follows inflammation — a rare condition in the mare, though common in the cow — more force will be requisite, and it may even be needful to incise the neck to the depth of one-sixth of an inch in four or more opposite directions prior to forcible dilation. The incision may be made with a probe-pointed knife, and should be done by a professional man if possible. The subsequent dilatation may be best effected by the slow expansion of sponge or seaweed tents inserted into the narrow canal. In such cases it is best to let the wounds of the neck heal before putting to horse."

114. Obstruction of Fallopian tubes resulting from excessive fatness. — In excessively fat animals, the Fallopian tubes may become mechanically obstructed by the pressure of fat tissue. This closure of the tube makes it impossible for the ova to descend into the uterus, and although the female may come regularly in heat and coition occur, the animal does not become pregnant. This condition does not necessarily involve fatty degeneration of the reproductive tissues, but may be associated with it. It is of course not possible to determine by external examination of the live animal whether failure to breed in a particular case is due to mechanical obstruction of the Fallopian tubes or to other causes. It is often possible to overcome this difficulty by dieting the animal. Very fat animals which do not breed should be placed upon a restricted diet which will cause them to lose weight regularly until they have regained a normal breeding condition.

115. Other causes of barrenness. — Other conditions which may be regarded as more or less temporary causes of barrenness are: (*a*) insufficient food supply, causing emaciation and a consequent failure of the sexual organs to mature ova; (*b*) failure of the animal to retain the semen of the male, due to unusual nervous irritability of the female sexual organs; and (*c*) sudden and marked change of condition, such as is brought about by the transportation of animals from one continent to another. (Plates V and VI.)

Some of the causes and treatment of sterility have already been somewhat fully discussed under the general subject of fertility. A mere mention of some of the above causes of sterility indicates the treatment. In many cases of sterility all that is needed is to remove the cause. The nervous irritability of the female sexual organs seems to be caused by, or at least associated with, an unusual flow of blood to the generative organs, causing congestion. This condition can be alleviated sometimes by exercising the female even to the point of exhaustion. It is a well-known fact that the Arabs were in the habit of riding their mares to exhaustion just before mating with the stallion. This treatment brought about a generally relaxed condition of the whole body and particularly of the reproductive organs, which is to be regarded as favorable for conception.

The importation of animals from foreign countries often results in temporary barrenness. This condition seldom or never becomes permanent.

116. Sterility from fatty degeneration. — The maintaining of animals in an excessively fat condition for a long period of time will eventually result in fatty degeneration of the tissues. When this condition attacks the ovaries, it frequently causes permanent sterility. Certain foods are believed to hasten fatty degeneration of the reproductive tissues. Tanner[1] holds that "this fatty degeneration of the ovaries has been traced to the use of foods rich in sugar. I have reason to believe that the action of sugar in its various forms is most important in its influence upon the generative system, and I think there is just cause for considering that any animal may by its use be rendered incompetent for propagating its species."

117. Sterility caused by abortion. — Among the important causes of infertility among the domestic animals probably none is responsible for so many failures to produce living offspring as abortion. Two kinds of abortion are recognized, non-contagious and contagious. Non-contagious abortion may result from a variety of causes closely associated with the environment of the animal. Law[2] gives a number of the more important causes of abortion in the mare. "The mare may abort by reason of almost any cause that very profoundly dis-

[1] Tanner, "The Reproductive Powers of the Domestic Animals," *Journal of the Royal Agricultural Society*, vol. 1, 1865, p. 267.

[2] Law, "Diseases of the Horse," U. S. Department of Agriculture, 1903.

turbs the system. Hence very violent inflammations of important internal organs (bowels, kidneys, bladder, lungs) may induce abortion. Profuse diarrhea, whether occurring from the reckless use of purgatives, the consumption of irritants in the food, or a simple indigestion is an effective cause. No less so is acute indigestion with evolution of gas in the intestines (bloating). The presence of stone in the kidneys, uterus, bladder or urethra may induce so much sympathetic disorder in the womb as to induce abortion. In exceptional cases wherein mares come in heat during gestation, service by the stallion may cause abortion. Blows or pressure on the abdomen, rapid driving or riding of the pregnant mare, especially if she is soft and out of condition from idleness, the brutal use of the spur or whip, and the jolting and straining of travel by rail or boat are prolific causes. Bleeding the pregnant mare, a painful surgical operation, and the throwing and constraint resorted to for an operation are other causes. Traveling on heavy muddy roads, slips and falls on ice, and jumping must be added.

"The stimulation of the abdominal organs by a full drink of iced water may precipitate a miscarriage, as may exposure to a cold rainstorm or a very cold night after a warm day. Irritant poisons that act on the urinary or generative organs, such as Spanish flies, rue, savin, tansy, cotton-root bark, ergot of rye or other grasses, the smut of maize and other grain, and various fungi in musty fodder are additional causes. Frosted food, indigestible food and, above all, green succulent vegetables in a frozen state, have proved effective factors, and filthy stagnant water is dangerous. Low condition in the dam and plethora have in opposite ways caused abortion, and hot, relaxing stables and lack of exercise strongly induce it.

The exhaustion of the sire by too frequent service, entailing debility of the offspring and disease of the fetus or of its envelopes, must be recognized as a further cause." The symptoms of abortion are similar to those of approaching parturition (see p. 75), if the threatened abortion occurs during the later stages of pregnancy. Abortion may occur during the first four weeks of pregnancy without any very marked symptoms. The fact of abortion is indicated by the animal again coming in heat. But, as already shown in cases of superfœtation, the occurrence of heat is not absolute evidence that abortion has resulted.

Ewart [1] has called attention to the fact that the mare is far more apt to abort at certain stages of gestation than at others. He regards the period from the sixth to the ninth week as one during which the mare is peculiarly susceptible to changes in her environment which may have a tendency to cause abortion. This is due to a change in the form of attachment of the fœtus to the uterus, from the primitive yolk sac to the more permanent villi. "At the end of the third week of gestation, when the reproductive system passes through one of its periods of general excitement, about one-fourth of the embryonic sac probably adheres to the uterus; but at the end of the sixth week, when another wave of disturbance arrives, all the grappling structures are at one pole. Hence, there is probably more chance of the embryo 'slipping' at the end of the sixth than at the end of the third week. About the end of the seventh week the supply of nourishment by means of the yolk sac is coming to an end, and there is, perhaps, still about this time an hereditary tendency for the embryo to escape. Unless the new and more permanent nutritive apparatus is provided, unless a

[1] Ewart, quoted by Marshall, *loc. cit.*, p. 615.

countless number of villi rapidly sprout out from the allantois, the embryo will die from starvation during the eighth week, and in a few days be discharged. It may, therefore, be taken for granted that there is a certain amount of danger at the end of the third and sixth weeks, but that the most critical period is about the end of the seventh or beginning of the eighth week; for unless the villi appear in time, and succeed in coming into sufficiently intimate relation with the uterine vessels, the developmental process is of necessity forever arrested."

Abortion among sheep seems to be largely due to debilitating conditions due to insufficient and unsuitable food, although Heape[1] has pointed out that shearling ewes are more apt to abort than those of maturer age.

118. Contagious abortion and sterility. — The most insidious, widespread and generally important cause of sterility, especially among cows, is due to a germ infection (*Bact. abortus*) which is recognized under the terms contagious or infectious abortion. It is found oftener in herds of dairy cattle than among beef breeds, not because dairy animals are more susceptible to this disease, but because they are generally handled in such a manner as to provide more favorable conditions for its spread. Beef breeds are generally less closely housed and they are more frequently permitted to calve on the pastures, thus avoiding two common circumstances favorable for the transmission of the disease. The infection is carried chiefly by the bull. If a healthy bull is permitted to serve a cow infected with the germ of abortion, he will generally transfer the infection to all cows which later may be served by him. The disease is not communicated to any important extent from cow to cow by merely

[1] Heape, "Abortion, Barrenness and Fertility in Sheep."

standing side by side in the same barn. The Scottish Committee appointed to investigate abortion found that infection might be carried from a diseased cow to a healthy animal by inserting a small wad of cotton into the vagina of a diseased cow for twenty minutes and transferring this to the vagina of a healthy pregnant cow or sheep. Such infection invariably caused abortion within a month. The specific organism *Bacterium abortus* is probably not the only germ which may cause abortion. MacFadyean [2] has called attention to the very great increase in the sterility of cows in Prussia and Switzerland during recent years. The cause of this sterility has been ascribed to a disease known as "infectious granular vaginitis." This affection produces an acute inflammation of the vulva and vagina and the infection is spread through a herd by the bull. Law [3] holds "that any micro-organism which can live in or on the living membrane of the womb producing a catarrhal inflammation, and which can be transferred from animal to animal without losing its vitality or potency is of necessity a cause of contagious abortion."

119. Treatment for contagious abortion. — The continuance of contagious abortion is generally prolonged even under the best of circumstances. Treatment may sometimes be very successful, and again any form of treatment may fail to make any lasting impression upon the disease. As the bull is the chief source of contagion, the treatment should start with him. Veterinarians recommend that the sheath and external genitals of the bull

---

[1] Law, "Diseases of Cattle," U. S. Department of Agriculture, 1908.

[2] MacFadyean, *Journal of the Royal Agricultural Society*, 1909, p. 337.

[3] Law, *loc. cit.*, p. 166.

be thoroughly disinfected with a solution of bichloride of mercury 1 to 2000 containing one per cent of copper sulfate. The same disinfectant should be used to flush the vagina of the suspected cows. Nocard [1] recommends the following solution for cleansing the external genitals, anus and tail of the aborting cow:

Distilled or pure rain water	2 gallons
Hydrochloric acid	2½ ounces
Corrosive sublimate	2½ drachms

These ingredients should be thoroughly mixed.

In Denmark, according to Dalrymple,[2] all membranes and the foetus are buried in lime and the internal genital organs of the cow are thoroughly disinfected with a one per cent solution of creolin or a half per cent solution of lysol. The external genitals of the bull and of cows about to be bred are treated with the same solution.

All aborted tissues must be burned and the premises thoroughly disinfected. This treatment is uncertain and often unsuccessful. A breeder will generally gain time in the eradication of abortion from his herd by insisting upon never using a contaminated male on young heifers bred for the first time or upon cows known to be free from the disease. All infected and aborting cows should be separated absolutely from the females which have never aborted. All cows known to be free from the disease should be covered by a bull also known to be free from infection. The uninfected herd must be guarded from infection from the diseased herd. This plan will eventually build up a clean herd and no other method so far devised is certain to accomplish this desirable result.

[1] Dalrymple, "Veterinary Obstetrics," p. 59.
[2] *Loc. cit.*

The problem of treating the bull and cows known to be infected is quite distinct from the building up of a clean herd. Fortunately this infection is not transmitted by heredity and is not necessarily spread from mother to offspring by direct infection. It is, therefore, possible for a breeder to retain in the young animals from the infected herd the best products of his skill and experience. When a cow from the infected herd aborts, all tissues expelled from the uterus should be promptly burned and the stall thoroughly disinfected by a generous use of lime. The vagina of the cow should be disinfected with the solution described on p. 125. The aborting cow should be permitted to rest at least six months before breeding again. The prevention of abortion in a cow already pregnant has been successfully accomplished in a number of instances by internal applications of carbolic acid. Taylor's[1] successful experiments in preventing impending abortion are worthy of note. A description of the treatment of one cow is typical and will serve as an example of the methods which were generally successful. A grade cow four years old that had aborted the previous year and was known to be infected with granular vaginitis was given the following treatment: At the beginning of the fourth month of the period of gestation she was given 200 cubic centimeters of a four per cent solution of carbolic acid in her feed. The dose was increased to 250 cubic centimeters the fifth month and to 300, 350 and 400 cubic centimeters for the sixth, seventh and eighth months respectively. This cow dropped a strong healthy calf at the end of the normal period of gestation. A summary of the results in one herd treated by Taylor shows

[1] Taylor, Montana Experiment Station, Bulletin No. 90, 1912.

that in 1908 there were fifteen per cent abortions, in 1909 twenty-five per cent, in 1910 five per cent, and in 1911 two-and-one-half per cent of abortions. The carbolic treatment was begun in December, 1909. It is recommended that infected males be treated in the same manner.[1]

120. Diagnosis of contagious abortion. — This disease is so widespread and causes such serious consequences when once well established in a herd that it is often of the greatest importance to be able to detect its presence in animals, even those which are not at the time pregnant. Various attempts have been made with greater or less success to secure a reliable diagnostic agent which would make it possible for the breeder to know which animals in his herd are infected with the germ of contagious abortion and thus be able to separate the infected individuals from those which are healthy.

The first substance of this character to be used was similar to tuberculin and mallein. The name of "abortin" was given to this substance by MacFadyean and Stockman. This material was to be injected into the circulation of the cows of a herd. The infected cows showed a considerable rise in temperature following the infection. The healthy cows showed but little or no increase in temperature. Brüll[2] at Vienna, after testing this method on a large number of cows, concluded that abortin was an unreliable diagnostic agent for determining the presence of contagious abortion in cattle. Other investigators have reported similar unfavorable results from the use of this material.

---

[1] *See also* Good, Kentucky Experiment Station, Bul. No. 165; Surface, Kentucky Experiment Station, Bul. No. 166; MacNeal and H. W. Mumford, Illinois Experiment Station, Bul. No. 152.

[2] Brüll, "Berl. Tierärtzt Woch.," Bd. 27, pp. 721–727.

In cases of typhoid fever, the so-called agglutination test was found to be a fairly reliable agent for the diagnosis of this disease. In 1907-8, Ginstead in Denmark applied this test successfully to cows suffering from contagious abortion. Later MacFadyean and Stockman [1] published the results of a limited number of investigations, reporting unfavorably upon this method. Later Holth and Wall,[2] after an extensive series of investigations involving hundreds of cows, concluded that the agglutination test was a reliable diagnostic agent but probably subject to larger error than the complement fixation test.

121. The complement fixation test.[3] — Neither the abortin nor the agglutination test has proven entirely satisfactory under all circumstances. The most reliable test now available is the complement fixation test.[4] Connaway of the Missouri Experiment Station says, "This test has been found very reliable as a diagnostic method in contagious abortion. The result of the test on some infected herds shows that in old infected herds the per cent of re-acting animals runs from 60 to 90 per cent." This is a highly complicated and difficult test to make, but will with practical certainty cause a reaction in cows that have been infected. Cows that have aborted may develop an immunity to this disease, and when this has occurred the complement fixation test cannot be used to distinguish between those cows which will abort

---

[1] MacFadyean and Stockman, "Report of the Departmental Committee to inquire into Epizootic Abortion," Pt. I, and Appendix, London, 1909.

[2] "Berl. Tierärtzt Woch.," Bd., pp. 686–688, 1909.

[3] *See* Surface, Kentucky Exp. Sta. Bul. No. 166; MacNeal and Mumford, Illinois Exp. Sta. Bul. No. 152; Russell, *Science*, N. S., vol. 34, p. 494, 1911; Wisconsin Research, Bul. No. 24, 1912.

[4] Missouri Experiment Station, Bul. 131, p. 486.

and those which are immune. The great value of this lies in the fact that the aborting cows may be entirely separated from the healthy members and eventually by this separation the disease may be entirely eliminated from the herd.

122. Sterility of free-martins. — The birth of twins among cattle is frequent. When a cow gives birth to twins, one a female and the other a male, the female is called a free-martin and is generally sterile. So far as known, this condition does not exist among any other known species of animal. Among sheep, for example, where twins are very common, the female twin born with a male may be even more fertile than the single born lamb. No case of sterility among human twins has ever been recorded where the sterile condition was believed to be due to the fact that one twin was a male and the other a female. Among cattle where both twins are of the same sex both are fully fertile. This is, therefore, a remarkable biological fact which it is difficult to explain. Morse[1] reports that Dr. Luer found 113 cases of twins, one a male and the other a female, in the records of the East Prussian Holland Herd Book. Of this number all the females were sterile except six.

The author has examined a considerable number of free-martins and in every case of sterility the female reproductive organs have been imperfectly developed. One case examined is typical. A grade Aberdeen Angus had every external appearance of a female. The location and form of the external genital organs was that of a true female. There were only two peculiarities which were visible externally. A tuft of long hair resembling the growth on the sheath of the bull grew from the lower

[1] Morse, *Breeder's Gazette*, vol. 64, p. 346.

extremity of the vulva. The mammary glands were very small and but little developed and resembled more the rudimentary mammary glands of a bull than the normal glands of a cow. At two years of age this supposed cow had never come in heat. She was permitted to run in the same paddock with two mature bulls for a period of six months, but never came in heat. This animal was slaughtered and the reproductive organs removed for examination. It was found that the animal had a vulva and a very short vagina-like organ ending in a blind sac. No uterus, Fallopian tubes or ovaries were found. There was also found a rudimentary penis. It seems probable that sterile free-martins are imperfect males or hermaphrodites. It is not possible at this time to give any satisfactory explanation of why there should exist a greater tendency to hermaphroditism when twins of different sex are born than twins of identical sex. It is still more difficult to explain why this phenomenon should occur exclusively among cattle. It is possible that the explanation of this phenomenon is associated in some way with the production of identical twins, but this does not offer any satisfactory explanation of why this strange occurrence should be found in the bovine species alone.

CHAPTER VII

HEREDITY

The characteristics of the individual are determined by heredity and development. What an animal may become, depends on the heritage received from its ancestors. No organic being can be developed beyond the limits imposed upon it by its inheritance. A favorable environment and good training will permit the individual to achieve the full limit of its possibilities, but no amount of training and no combination of favorable circumstances can ever lift the individual above the inheritance which it has gained through its parents. The trotting horse may have inherited the capacity to trot a mile in two minutes, but if its development has been arrested by insufficient food and an unfavorable climate, and no attempt has been made to develop its inherent ability to go fast, it can never achieve the full measure of its possibilities. It is also true that if a horse has not inherited from a line of trotting ancestors the ability to go fast at the trot, no amount of training and no system of feeding can develop the animal to a point where it will be able to trot a mile in two minutes.

Heredity then represents what an animal really is or can become. The individual cannot in any manner or to any extent influence its own heredity. An animal's inheritance is determined by its ancestors. The individual may profoundly influence its development through seeking a favorable environment and through habitual use or disuse of its inherited tendencies.

123. Development. — The full realization of the inherited capacities of an animal is accomplished through its environment and training. The most important factors concerned in the environment of an animal are food and climate, and these acting upon the inherited qualities of the animal may profoundly influence the actual characteristics of the individual. In order that the inherent characteristics of any organic being may attain to full development, a sufficient supply of food must be available. When food is scarce, the individual may be unable to develop, and what it may ultimately become, may be greatly influenced by this lack of food. A full and sufficient food supply may cause the individual to develop beyond the average condition of the species, particularly if the average environment does not furnish a generous supply of food.

The training of the individual, likewise, is an important factor in determining its ultimate development. Any function of an animal which is not exercised may retrograde or in some cases practically disappear. The milking function in domestic cattle is a good example of the influence of both development and exercise or training. The highly developed dairy breeds of cattle, when generously fed and carefully milked, produce very much more milk than their wild ancestors. Individual cows of modern dairy breeds, if starved and carelessly milked or neglected, will fail to develop the highly specialized milking function.

124. Heredity defined. — In much of the literature of biology pertaining to heredity, there is a lack of definite-

HEREDITY

ness in the use of terms. The literature of animal-breeding is still less exact in its terminology. It seems important, therefore, that in the very beginning we should have as clear a conception as possible of the definitions of heredity which have been proposed. Heredity is the organic relation existing between an individual and its ancestors. It is the continuous biological thread connecting generations of organic beings. Heredity is "organic resemblance based on descent."[1] Thomson[2] has defined heredity as "the organic or genetic relation between successive generations." "Understood in its entirety," says Herbert Spencer,[3] "the law is that each plant or animal, if it reproduces, gives origin to others like itself; the likeness consisting, not so much in the repetition of individual traits as in the assumption of the same general structure." "The transference of similar characters from one generation of organisms to another, a process effected by means of the germ-cells or gametes."[4] "By inheritance," says Lock,[5] "we mean those methods and processes by which the constitution and characteristics of an animal or plant are handed on to its offspring, this transmission of characters being, of course, associated with the fact that the offspring is developed by the processes of growth out of a small fragment detached from the parent organism." Another definition of heredity is that it is the tendency of the offspring to be like the parents. There exists a definite organic resemblance or

[1] Castle, "Heredity in Relation to Evolution and Animal Breeding."
[2] Thomson, "Heredity" (1908), p. 13.
[3] Herbert Spencer, "Principles of Biology," vol. I, p. 301.
[4] R. H. Lock, "Recent Progress in the Study of Variation, Heredity and Evolution," 1906, p. 292.
[5] *Loc. cit.*, p. 1.

relation between parents and offspring. This relation is a universal biological phenomenon and is called heredity. There exists a continuous line of descent from generation to generation. The mechanism of heredity is to be found in the protoplasm of the germ-cells. The stream of descent is from germ-cell to germ-cell. The soma- or body-cells constitute in a sense only a temporary abiding place for the germ-plasm.

125. Heredity and variation not antagonistic. — It is not stating the facts correctly to maintain that the tendency of offspring to be like the parent, which we call heredity, is opposed by the universal tendency of organisms to vary. These are but two phases of the same phenomenon. "Living beings do not exhibit unity and diversity," says Brooks, " but unity in diversity. These are not two facts but one. The fact is the individuality in kinship of living beings. Inheritance and variation are not two things but two imperfect views of a single process." There is a sense, of course, in which variation is opposed to heredity. It is conceivable that recombinations of characters may occur in the germinal substance, and these new combinations may cause such modification of characters already present that the new organism may be radically changed.

Heredity is the genetic relation of parents and offspring. On the average the offspring will be like the parents. But this relation admits of variations within more or less definitely prescribed limits. The offspring have an individuality all their own, but this does not preclude the existence of a genetic continuity which is the common heritage of parents and offspring.

126. The kinds of heredity. — Every individual is the result of a union of the germ-plasm of two individuals.

The evidences of this dual origin are not always exhibited in the same manner. In some individual offspring the characters of one parent may predominate, in others the parental characters seem to blend so successfully that often the child differs from either parent, while in still others the characters of both parents appear in the offspring with apparently equal force but, instead of blending, the characteristics of the parents appear clearly as distinct and easily recognizable qualities. These methods of transmission were called by Galton, blending, alternate and particulate inheritance.

127. Blending inheritance. — In this type of inheritance the characters of the offspring represent a blending or intermingling of the characters of the two parents. Stature in man may and often does represent the blending of the statures of the two parents. The stature of the son or daughter is taller than the shorter parent, but falls short of that of the taller parent. When a relatively small mare is mated with a heavy draft stallion, the resulting offspring is never so large as the sire nor so small as the dam, but represents an approach to a mean between the two. Another example of blended inheritance is to be observed in the cross-bred lambs resulting from the union of a sire of the coarse wool type and a dam belonging to the fine wool or Merino breed. The lambs of such a cross are covered with wool which in respect to density, length of staple and fineness of fiber represents a blending of the characters of the parents.

In some cases of supposed blending inheritance, the characters have been observed to follow the law of dominance and segregation discovered by Mendel.

128. Alternative inheritance. — A character may be transmitted intact from one parent directly to the offspring without apparent modification by the other parent. In this type of inheritance one parent seems to possess a predominating influence in determining the characteristics of the offspring. Many examples of this type of heredity are to be observed among the domestic animals. The white face of the Hereford breed of cattle is invariably transmitted, even when one parent belongs to a widely different breed. The quality of speed in horses is undoubtedly transmitted, sometimes in accordance with Galton's alternative inheritance. Fecundity in the domestic fowl has been shown by Pearl to be transmitted through the male. The hen inherits fecundity directly from the sire. The ability to lay a large number of eggs is not transmitted from the mother to the immediate female offspring, but to her male offspring.

"Hence if the daughters of high producing hens are selected, one does not get in them the high productiveness of the mother. It is her sons that inherit the character, although they cannot show it except in their offspring."[1]

129. Particulate or mosaic inheritance. — The characters of the parents are often transmitted in such a manner that they are not in any sense blended but appear rather as a mosaic. The color character is frequently inherited as a mosaic. A very good example of this kind of inheritance is seen in the Holstein Friesian breed of cattle. This breed is black and white, these colors appearing in definite, clearly defined areas and not blending. The Holstein Friesian breed was originated by crossing a black breed and a white breed of cattle. The colors black and white in other animals seem to behave in a similar manner. When white hogs are mated with black, the offspring are always spotted. Recently it

[1] Morgan, "Heredity and Sex," p. 213.

has been suggested that particulate inheritance is in reality true alternative inheritance in which the mosaic result is caused by the absence of the factor for uniformity.[1]

130. Mendelian inheritance. — Progress in the investigation of breeding problems has come through statistical investigations, by cytological studies of the germ-cells themselves, and by experimental breeding. Each of these methods has contributed evidence of value in the direction of a more definite understanding of the principles of heredity. In recent years experimental breeding has contributed very materially to our knowledge of the science of heredity. Improvements in the technique of cell studies has supplemented the results obtained by experimental breeding. It is a significant fact that the hypotheses based upon experimental breeding agree in many important particulars with those derived from minute investigations of the origin and development of the germ-cells.

Perhaps the most remarkable series of investigations in experimental breeding were those carried on by Johann Gregor Mendel, an Austrian monk. These interesting investigations gave a new impetus to the study of theoretical heredity and particularly to the practical improvement of plants and animals. The investigations of Mendel were first published in 1865 and over twenty years later were again published in the Transactions of the Natural History Society of Brünn. No attention was given to this important contribution to the science of breeding, and even Nägeli, a former teacher of Mendel, to whom the results were submitted, failed to recognize their fundamental significance. It was not until the year 1900

[1] Walter, " Genetics," p. 164.

that three great investigators, De Vries in Holland, Von Tschermak in Austria, and Correns in Germany, simultaneously discovered the published results of Mendel and recognized their great fundamental importance.

131. The experiments of Mendel. — Mendel selected for his experiments the common garden pea. It is not certain that he fully recognized the wisdom of making such selections as were finally made for his work, but at any rate he selected two varieties of peas differing in a simple character but each firmly fixed in the parent variety. The peas were also self-fertilizing, and accidental mixture by cross fertilization was thus avoided. The characters selected by Mendel were " purple or white flowers," " yellow or green cotyledons," and " round or wrinkled seeds." He found that all of these characters were firmly fixed and bred true. These varieties were crossed and the progeny were bred pure for many succeeding generations.

He decided upon the simple character of color and selected a green seeded and yellow seeded variety. Reciprocal crosses were made, and from each cross the resulting peas were all yellow. Because the yellow color in the cross appeared in every case, he called it the dominant character; and the green color which did not appear in the first generation was called a recessive character.

All the yellow seeds of the first generation resulting from the original cross were planted. In the second generation Mendel discovered that both green and yellow seeds appeared. In calculating the relative proportion of the two colors, he found that about one-fourth of the seeds were green and the remaining three-fourths yellow. The green seeds and yellow seeds were planted separately and it was found that the green seeds produced only green seeds. They were planted for several generations and always came true, showing no yellow character. When the yellow seeds were planted, however, Mendel found that a certain proportion of the yellow seeds had both green and yellow offspring and a certain proportion had only yellow offspring. The latter remained fixed and true in character when bred for several generations.

The results of these investigations carried through many generations indicated that there was a certain mathematical ratio traceable in the offspring resulting from the crossing of these two distinct varieties of peas. In these results Mendel found that in the first generation the dominant character (yellow seed) appeared to the exclusion of the green color. In the second generation he found that 25 per cent of the offspring were green and 75 per cent apparently yellow. If the green peas were planted, they produced only green peas, but when the yellow peas were planted, they produced 25 per cent pure yellow, 50 per cent mixed or hybrid (yellow and green), and 25 per cent pure green. The pure yellows and pure greens continued to breed true, but the 50 per cent " hybrid " peas continued to split up in each generation in the proportions of 25 per cent pure green, 25 per cent pure yellow, and 50 per cent hybrid.

132. The law of dominance. — From these results Mendel formulated the law of dominance, which is that when two contrasting characters are bred together the offspring in the first (F_1) generation will all exhibit the dominant character.

133. The law of segregation. — When the individuals comprising the first generation are interbred, the resulting offspring (F_2 generation) will possess the characters in the proportion of three of the dominant character to one of the recessive. The recessive character is pure and will breed true. The individuals possessing the dominant character will be made up of one-third pure dominants and two-thirds hybrid dominants in which the recessive

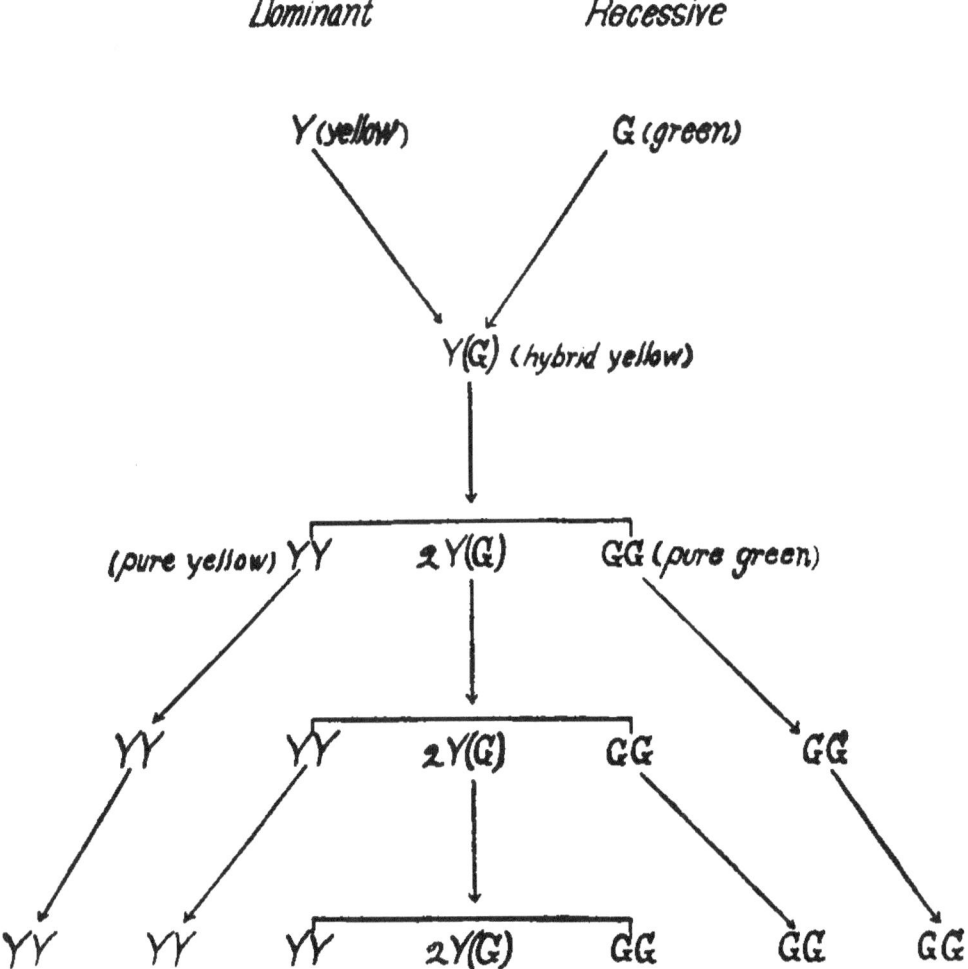

Fig. 16. — Diagram illustrating mendelian inheritance of yellow and green characters in the garden pea.

character will reappear in the next generation. This may be more clearly shown in the preceding diagram (Fig. 16).

Mendel's hypothesis affirms that when animals or plants of contrasted characters are bred together, these characters do not blend in the germ-cells of the offspring.

HEREDITY

Fifty per cent of the germ-cells will contain the dominant character and fifty per cent the recessive character. This is true of both male and female germ-cells.

134. Unit characters. — Mendel's theory presupposes the existence of unit characters, so-called because they are transmitted as independent units. There exist in every organic being a very large number of unit characters, and these may be determined by experimental breeding. Up to the present time a relatively small number of unit characters have been definitely differentiated and described, but our knowledge in this direction is being rapidly extended.

135. Gametic purity. — In one stage of maturation of the male and female germ-cells, the nucleus of each contains but half the normal number of chromosomes characteristic of the species. This is the final stage in the maturation process before fertilization takes place. The germ-cells in this stage are called gametes. The fertilized egg-cell which results from the union of a mature sperm and egg-cell (gametes) is called a zygote (fertilized egg-cell). The zygote, therefore, contains the normal number of chromosomes characteristic of the species, one-half derived from the egg and one-half from the sperm-cell. Gametic purity is a term used to designate the discontinuous nature of unit characters. The gamete in Mendel's pea contains the factor necessary for the production of yellow seeds, or it does not.[1]

The terms homozygous and heterozygous were proposed by Bateson to designate the fundamental constitution of the germ-cells in respect to inherited characters. " An individual is said to be homozygous for a given

[1] Darbishire, "Breeding and the Mendelian Discovery," p. 217.

character when it has been formed by two gametes each bearing the character, and all the gametes of a homozygote bear the character in respect of which it is homozygous. When, however, the zygote is formed by two gametes of which one bears the given character while the other does not, it is said to be heterozygous for the character in question, and only half the gametes produced by such a heterozygote bear the character. An individual may be homozygous for one or more characters, and at the same time may be heterozygous for others."[1]

The conception of "gametic purity" as originally stated requires some modification. It is no longer held that characters are transmitted as units, but rather the factors which combine to form characters. The factors are so far purely imaginary.

When mendelism was first seriously considered, there was no doubt among its most enthusiastic exponents that the characters existed in a pure unmixed state in the gamete.[2]

136. Application of Mendel's law. — Can the practical breeder apply the principles of heredity embodied in Mendel's law to the improvement of the domestic animals? The domestic animals are valued by man because of certain desirable characteristics which they possess. These characteristics are clearly recognized by the breeder. The meat animal is produced because it is endowed with certain qualities which give it a special value either to the consumer or the producer. The dairy cow is highly prized because she has the ability of producing large quantities of milk, cream or cheese. The horse is the burden-bearer. Its value depends on the amount

[1] Punnett, "Mendelism," 1913, p. 28.
[2] *See* Castle, "General Heredity," vol. V, No. 3, p. 93.

of energy it can develop either in tractive power for pulling heavy loads or in the form of speed or graceful action for the pleasure of the owner. Other animals are maintained in a state of domestication for other values of various kinds which contribute to the food, clothing or pleasure of man. In a sense the producer of live-stock may be compared to a manufacturer who employs capital, labor, raw materials and efficient machines for the production of more desirable and concentrated products. In this comparison the raw materials are the grain, hay and grass; and the efficient machine is the animal. The farmer's success and the interests of the consumer as well are greatly dependent upon the efficiency of this animal machine. Can Mendel's law be utilized in the efforts of the breeder to increase the efficiency of animals? If so, what does the breeder need to know in order to utilize the law of Mendel in further enhancing the value of the prevailing types of domestic animals?

137. The complexity of animal characters. — It must be recognized in the beginning that the qualities which are commonly mentioned by the breeder as highly desirable are generally the result of a combination of many characters. These combinations do not behave as simple unit characters. One of the first steps in the application of Mendel's hypothesis to practical breeding must be to analyze the valuable qualities of animals and determine as far as possible the unit characters. It is probably true that some combinations will behave in transmission in the same manner as simple unit characters. But if there are complex characters which behave in this manner they must be determined by careful investigation. When the qualities of an animal have been differentiated and their behavior in transmission determined, then the practical breeder may base his breeding methods upon the principles of segregation and dominance which are foundation stones in the theory of mendelian inheritance.

It will also be important to remember that to utilize the mendelian principle in the breeding of animals, the breeder must be dealing with contrasting characters. In the larger number of cases, the useful qualities of the highly improved animals of the farm are but modifications of characters which were already present in the wild forms. The function of milk secretion is common to all mammals. The domestic cow is valuable because this function has, through selection and skillful mating, been gradually improved. The quality of giving a small quantity of milk found in the wild cow, and the quality of giving a large quantity of milk as present in the highly improved dairy cow, are not contrasting characters. The one is but a modification of the other and is probably the result of accumulated fluctuating variations which have been preserved by methodical selection. It will be interesting to note here some characters among animals which have already been found to conform to Mendel's law.

138. The inheritance of polled and horned character in cattle. — Examples of Mendel's law are much more frequent among plants than animals. The complicated nature of animal characteristics has made it difficult to trace the workings of Mendel's law so far as it is related to many animal characters. It is also more difficult to find contrasting characters. An exception to the above must be noted in the case of the horned and polled characters in cattle. Whenever a pure polled animal is mated with a pure horned animal, the offspring in the first generation are all polled. If the first generation offspring are interbred, the polled and horned characters separate in the second generation in the proportion of 25 per cent pure polled, 50 per cent hybrid polled, and 25 per cent pure horned. The pure polled individuals,

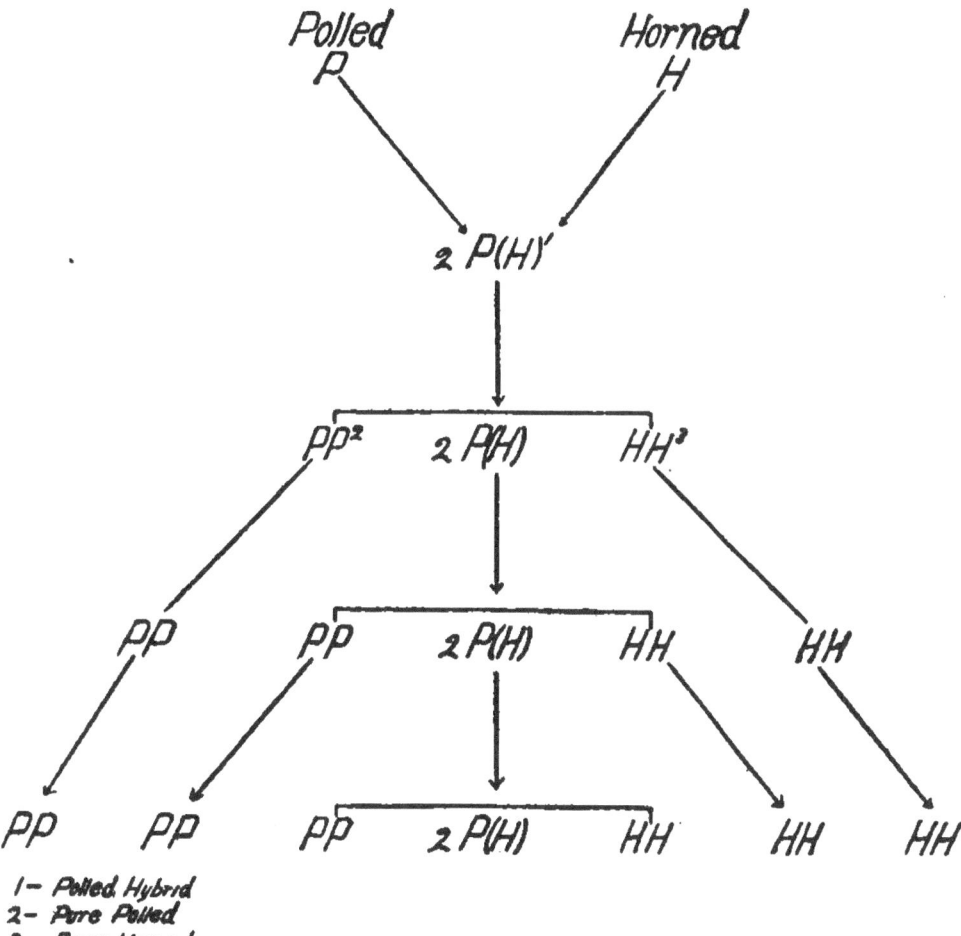

Fig. 17. — Diagram illustrating mendelian inheritance as exhibited in the transmission of polled and horned characters in domestic cattle.

if bred to others of their own kind, produce only pure polled offspring. If the pure horned offspring likewise are mated with pure horned individuals, their offspring produce pure horned offspring. If the 50 per cent hybrid polled individuals are bred to other hybrid animals, the result is the same as in the second generation. The offspring will be 25 per cent pure polled, 25 per cent pure horned, and 50 per cent hybrid polled. (See Fig. 17.)

This principle has been repeatedly used in the production of polled breeds of cattle. The only practical difficulty is encountered in determining which of the second generation are pure polled and which are hybrid polled. This difficulty can be overcome only by repeated matings and by observation of the offspring.

139. Theory of pure lines. — The practical breeder of animals has depended almost entirely upon methodical selection for the improvement of domestic animals. It was Darwin's opinion that the improvement accomplished in the desirable qualities of animals and plants was due to the persistent selection of desirable continuous variations. In bringing about improvement, therefore, the breeder only required keen powers of observation to detect any variations in a standard sort which were better than the qualities of the ancestors. By selecting these varieties and continuing this process for many generations, highly improved sorts were ultimately developed which came true when bred together.

In the practical application of this theory, it has been frequently discovered that the limits of improvement through selection were quickly reached. Apparently the degree of improvement could not be carried beyond a certain definite point. In a more careful analysis of the fundamental basis of improvement by selection, Johanssen[1] has demonstrated that very many of the domestic plants are not possessed of single characters only, but are a mixture. The selection exercised by man in such cases is essentially a process of selecting out

[1] Johanssen, 1909, "Elemente der exakten Erblichkeitslehre."

HEREDITY

the most desirable strain and gradually eliminating the less desirable. This is the pure line theory and is now generally accepted as applying to many cases of improvement, especially among plants.

An acceptance of this theory recognizes the fact that no amount of selection can improve a pure line after it has been separated by continuous selection.

An hypothetical example of pure line selection among animals might be imagined in the case of the wool of sheep. The wool produced under a given set of environmental conditions in a flock of sheep might vary from eight to twelve pounds. In the germ-cells of a given individual, we may assume that determiners are present for the production of eight, ten and twelve pounds. In breeding, these varying tendencies may be separated. It is conceivable that some of the offspring may have inherited the tendency to produce twelve pounds of wool, while others may have inherited the tendency to produce eight pounds. Through many generations of intelligent selection, the flock-master may bring about a more or less complete separation of the tendency to produce twelve pounds of wool and may thus increase the average production of wool from his flock. The application of this theory to animal-breeding is more difficult than to self-fertilizing plants, but the difficulties are partially removed by close interbreeding.

140. Hallett's wheat-breeding. — The pure line method of breeding probably explains Hallett's unusual success in the improvement of wheat in Great Britain. However, it must be said that Hallett believed that improvement within a pure line of selection was possible. Hallett's method may be best described by using his own words:

" A grain produces a ' stool ' consisting of many ears.

I plant the grains from these ears in such a manner that each ear occupies a row by itself. . . . At harvest, after the most careful study and comparison of the stools from all these grains, I select the finest one, which I accept as proof that its parent grain was the best of all, under the peculiar circumstances of that season. This process is repeated annually, starting each year with the proved best grain, although the verification of this superiority is not obtained until the following harvest.

"During these investigations no single circumstance has struck me as more forcibly illustrating the necessity of repeated selection than the fact, that of the grains in the same ear one is found greatly to excel the others in vital power."[1]

Hallett's experience has demonstrated, first, that great improvement may be secured by this method of selection. Second, he advocated the "ear to row" method. Third, it is probable that the results may be satisfactorily explained by Johanssen's pure line theory.

141. The presence and absence hypothesis. — In Mendel's conception of the theory of dominant and recessive characters, there existed a definite determiner for both the dominant and recessive characters. Each character appeared in the gametes in a definite form. Later investigations have pointed to the fact that the dominant character may be due to the presence of a specific determiner, while the recessive character may be due to its absence. This conception of the behavior of the mendelian pair of characters is quite different from Mendel's explanation.

The presence and absence theory has contributed materially to the science of genetics. This hypothesis

[1] *Journal of the Royal Agricultural Society*, vol. 22, p. 371.

has made it possible to harmonize many of the observed phenomena with the mendelian principle. It has also directed the attention of investigators to the fundamental nature of the characters themselves. Great progress will undoubtedly be made in the future in the direction of a more thorough study and consequently better understanding of the real nature of the characters of both plants and animals.

In the investigations on the eye color of man, it has been found that the dominant character is due to a brown pigment, while the recessive character is the result of the absence of this pigment. Darbishire [1] has clearly indicated the application of this theory in the case of the round pea and the wrinkled pea. The quality of roundness or wrinkledness as found in the garden pea is due to a difference in the amount of starch in the pea. In the case of the wrinkled pea, all of the sugar content is converted into starch. In the round pea a much more complete transfer of sugar to starch is accomplished. The inference is that in the round pea there is present something in the germ-cell that so affects the physiological processes in the cell that the sugar is more or less completely turned to starch. The something which determines this change of sugar to starch in the round pea is absent from the germ-cell in the wrinkled pea. Therefore, we have two varieties of pea that are clearly different. When the wrinkled pea and the round pea are crossed, they behave in accordance with the mendelian law. It is not necessary in this case to assume that the determiner which ultimately results in a round pea is entirely absent in the wrinkled pea, but it seems to be entirely correct

[1] Darbishire, "Breeding and the Mendelian Discovery," p. 127.

to assume that there exists an insufficient quantity of this substance and hence an incomplete transposition of sugar to starch in the wrinkled pea.

It is not at present possible to apply the presence and absence hypothesis to all cases of apparent mendelian inheritance. That it applies to a very large number of characters is obvious, but as pointed out by Darbishire there are many cases which cannot at the present time be explained on this theory, and in fact will be obstacles in the way of its general application. Some characters seem to be dominant in one class of plants or animals and recessive in another. The polled character in cattle is unquestionably dominant, while the possession of horns is a recessive. In the case of sheep the reverse seems to be true. The white color of pigs is dominant to black, but the black color of sheep is dominant to white.[1]

142. The theory of mutations. — It is clear that all improvements in the domestic animals must come through variation. If the offspring was always an exact reproduction of the parent, improvement would be impossible. But how have the improved qualities now possessed by our domestic animals come about? Have these qualities come through a gradual and continuous series of changes, each better than the last, or have they come through sudden and radical variations? A study of the ancestral history of both plants and animals gives clear evidence that new types have originated by both kinds of variation.

143. Two important classes of variation. — Darwin recognized these two distinct types of variation, but believed that the most important changes in organic beings were due to small but gradual and continuous variations in a given direction. In his earlier writings

---

[1] Punnett, "Mendelism," p. 29.

Darwin emphasized the great importance of continuous or fluctuating variations. He says, " It may be doubted whether sudden and considerable deviations of structure such as we occasionally see in our domestic productions, more especially with plants, are ever permanently propagated in a state of nature. Almost every part of every organic being is so beautifully related to its complex conditions of life that it seems as improbable that any part should have been suddenly produced perfect, as that a complex machine should have been invented by man in a perfect state."

Sudden variations were called by Darwin discontinuous. Variations frequently occur which vary widely from the parent form. Examples of this kind of variation are common in both plants and animals. The normal fruit of a peach tree is a typical peach having a rough skin and a flavor and color peculiar to the peach. Branches of peach trees, however, sometimes produce a fruit known as a nectarine. This is smooth, smaller than the peach, and possessed of a color, flavor and physical consistence markedly different from the normal fruit of the peach.

Among animals mutations frequently occur. Variations in the number of digits have been recorded by a large number of investigators. Huxley describes the case of a man born with six fingers on each hand and six toes on each foot. Four children were born to this man. The first, a male child, was born with six fingers on each hand and six toes on each foot. The second child, a boy, had five fingers and five toes, but one toe was deformed. The third child, also a boy, had five perfect fingers and toes, but the fourth child, a girl, although having the normal number of digits, had deformed thumbs.

The mutation appearing in the father was not only a radical variation, but there existed a strong tendency to transmit the abnormality. The strength with which mutations are transmitted is still further indicated by the later history of the sons and daughters of the father. "Salvator had four children — they were two boys, a girl, and another boy — the first two boys and the girl were six-fingered and six-toed like their grandfather; the fourth boy had only five fingers and toes.

"George had only four children; there were two girls with six fingers and six toes; there was one girl with six fingers and five toes on the right side, and five fingers and five toes on the left side, so that she was half-and-half. The third, Andre, you will recollect, was perfectly well formed, and he had many children whose hands and feet were all regularly developed.

"Marie, the last, who of course married a man who had only five fingers, had four children: the first, a boy, was born with six toes, but the other three were normal."[1]

Some marked variations among the domestic animals which have been recorded from time to time and which are probably true mutations may be mentioned. The normal foot of domestic swine is cleft, but solid-hoofed hogs are common, and this mutation is strongly transmitted. The breed of swine known as "mule-footed hogs" is an example. Horned sheep normally possessing two horns are sometimes born with four horns. The offspring of horned cattle are sometimes born without horns, and this variation is strongly transmitted.

To Hugo De Vries of Amsterdam we owe our present definite notions regarding the theory of mutations. While such variations were recognized by Darwin and other

[1] Huxley, quoted in Miles' "Stock Breeding," p. 79.

investigators, it remained for De Vries to demonstrate by scientific investigation the important rôle played by mutations in the evolution of plants and animals. This theory undertakes to explain the origin of species by sudden marked variations rather than by continuous or fluctuating variations. De Vries' experiments were successful not only in demonstrating clearly the remarkable tendency of certain species to vary abruptly, but of even more significance was his demonstration of the fact that such variations were as surely transmitted as were the regular or normal characters of the species.

It is true, mutations must not be confounded with the appearance of freaks or monstrosities among plants and animals. Often through arrested development or accidental injury during the embryonic existence of animals, certain characteristics may become so modified that they appear in the form of new characters. Such freaks are not transmitted and are not therefore mutations. The term mutations is used only to designate those variations which are heritable.

144. Kinds of mutations. — Mutations may be additions or improvements to the life of organic beings, or they may result in diminishing the ability of an animal or plant to live and thrive in an ordinary environment. De Vries has therefore suggested a classification of mutations as progressive, regressive and degressive.

Progressive mutations are those which have contributed something entirely new. It is in reality an additional character. Attention has already been called to cases of variation in which the offspring is provided with extra fingers or toes. An example of this form of mutation is described by Alexander Graham Bell in the case of multi-nippled sheep.

In regressive mutations the animal actually loses some hereditary character that it has formerly possessed. An animal may be born with less than the normal number of digits, and this may be transmitted to offspring. The so-called tailless breed of cats is probably descended from an individual born without a tail. A polled animal, the offspring of horned parents, would also be an example of regressive mutation.

The term degressive has been applied to such variations as have existed in the previous ancestral history of the animal or plant. When such variations reappear after many generations, they are known as degressive mutations.

145. Importance of the mutation theory. — The importance of the mutation theory to the animal-breeder lies in the fact that many of the most important and useful qualities present in the domestic animals have probably arisen through sudden variations or mutations. Such mutations are likely to occur at any time. It is indeed probable that such mutations do occur frequently. The intelligent breeder who understands the laws of evolution will clearly recognize the importance of such mutations. The fact that they are certainly transmitted by heredity has given to the breeder a most important method in the permanent improvement of the domestic animals. It is, of course, quite as true that mutations may be degressive and thus actually be of less value than the parent forms. Certain breeds are probably more variable than others, and this fact may be both an advantage and a disadvantage. It is a desirable condition in that mutations may occur and some of these may be improvements over the parent species. Such a tendency to vary may be a disadvantage in that the variations may sometimes be in the direction of less valuable characters. Eventually no breed of animals can be most useful until the desirable characters are firmly fixed and reasonably certain to be transmitted by inheritance.

146. Mono-hybrids and di-hybrids. — In all of the examples so far used to illustrate the principles of segregation and dominance as developed by Mendel, only such simple unit characters have been employed as indicate clearly the law of Mendel. Such simple contrasting characters are known as mono-hybrid. It is easy to demonstrate the mendelian hypothesis in the case of mono-hybrids.

But among domestic animals the qualities which have made organic beings useful to man are in most part a combination of two or more unit characters. In such cases it is far more difficult to use the mendelian formula. Mendel himself determined the probable application of the principle to cases where two different unit characters were present. He crossed the wrinkled green peas with smooth yellow peas. Manifestly, the number of combinations of characters was greater than when mono-hybrids were examined. The proportion of 3 plus 1 is the universal result in the F_1 generation where single unit characters are involved. Mendel found in the case of combinations of two unit characters that the mathematical statement (3 plus 1)^2 (16) represented the true result. There would appear sixteen possible zygotes as a result of crossing individuals containing two unit characters in each of the parents. In the offspring resulting from such crosses the characters would be combined in such a manner as to produce sixteen kinds of individuals.

It is apparent that if the number of combinations of

unit characters increases, the number of possible different individuals produced by crossing similarly increases. In the case of tri-hybrids in which three characters combine, the number of different individuals or zygotes produced would be represented by (3 plus 1)^4. For the purposes of the practical breeder, therefore, little progress is possible except in cases where one or two characters only are involved. In those cases in which the characters involved are made of a combination of many units it is first necessary that the important unit characters be separated by long-continued breeding; in other words, until it becomes homozygous. By such methods, it is possible to obtain recombinations, and from the point of view of the practical breeder, entirely new characters. The mendelian principle may, therefore, in this way become a powerful instrument in the future improvement of domestic plants and animals.

CHAPTER VIII

INHERITANCE OF ACQUIRED CHARACTERS

THE normal characteristics of all organic beings may be greatly modified by external conditions acting continuously upon the organism for a longer or shorter period of time. Such modifications may be of so marked a character as to cause an apparent departure from the average or normal characteristics of the race. Through accident, disease or by design, the animal or plant may become permanently mutilated. The modifications which result from the causes mentioned may make the domestic animal or plant more useful to man than the normal organism; the changes may in fact be real improvements, and as such it would be very desirable to perpetuate them by heredity. Thus the breeder of domestic animals, having accurately observed the general fact of inheritance of the normal characters, has reasoned that such remarkable changes as those often resulting from use or disuse, favorable environment or mutilations must be transmitted with equal force.

147. Belief in transmission of acquired characters. — The literature of the ancients indicates that the philosophers of that period believed in the inheritance of acquired characters. Aristotle mentions the transmission of the exact shape of a cautery mark. At a much later period Lamarck elaborated his theory of variation and selection which took for granted that all modifications resulting from use and disuse and the effects of environment were transmitted by heredity.[1]

Among those who have failed to find satisfactory evidence of the inheritance of modifications must be mentioned Kant, Blumenbach, and later Galton, Weismann, Ray Lankester and practically all the leading biologists of modern times.

148. Practical breeders believe in transmission of acquired characters. — There is a widespread belief among the breeders of domestic animals that acquired characters are inherited. To the practical breeder of dogs who has observed the sensitive reaction of the fox hound to the scent of the fox, or the alertness of the setter or pointer in the presence of a fresh bird track, it is difficult to find a satisfactory explanation for existing facts without assuming that the results of the training of a dog to some extent must be transmitted. Extreme speed in running and trotting horses is the result of training and exercise. The offspring of parents who have acquired extreme speed through training and exercise are more apt to possess the ability to acquire similar extreme speed than the offspring of parents who have not been trained. The practical breeder concludes, therefore, that the acquired speed must be transmitted.

There are many breeders of beef cattle who firmly believe that by maintaining the breeding animals in high condition, the calves will have a more pronounced development of those characters which are recognized as belonging to the beef type than will the calves of parents which are maintained on a low plane of nutrition.

The breeder is an accurate observer, and there can be no question that his facts are generally correct in this

[1] Thomson, "Heredity," p. 170.

instance as in many others. But the biologist could easily explain the facts cited upon wholly different grounds. The fact that the trotting horse parents could be developed to high speed indicated clearly that they had inherited a capacity for such development. The offspring inherited the same capacity and under the same favorable conditions would develop the same speed. The breeder of beef cattle is in general correct in his practice when he maintains his breeding animals on a high plane of nutrition, not because such treatment in any way influences the fundamental constitution of the germ-plasm, but because this is a satisfactory method of determining which breeding animals possess the fattening tendency. Those parents that show the capacity for rapid and economical fattening are preserved and the unthrifty are eliminated from the breeding herd. The high plane of feeding is a method of selection.

149. Nature and nurture. — The nature of an animal is determined by heredity. What an animal may become is limited by its inheritance. What an animal actually becomes is often largely determined by the use made of its inheritance. The individual may inherit great possibilities, but these may never be realized because undeveloped. The characteristics of an individual, then, are the result of both nature (heredity) and nurture (development).

The tremendous importance of nurture to the individual is demonstrated by numerous examples. The untrained trotting horse may have inherited the same possibilities for development as his stable mate, but because of superior opportunity through skillful training the latter makes a new trotting record while the former achieves nothing.

The most notable examples of acquired development are to be found in the human family. Through the lifetime of a man the mental and physical qualities may be greatly modified. An individual may acquire great mental power. Such acquirement may have been achieved under very great difficulties. The particular kind of mental efficiency represented by the development of such an individual may be an exceedingly desirable characteristic of the greatest value to the human race. Its transmission by heredity would be desirable for the good of the race. Can such acquired characters or habits be transmitted? The answer to this question is important to the development of the human race. It is likewise of the greatest economic importance to the breeder of domestic animals. In the domestic animals the highly artificial characters possessed by the improved forms of cattle, horses, sheep and swine are very largely due to the development or good handling to which these animals have been subjected. If the results of the high degree of development of one generation are to fundamentally influence the characters of the next, then a new significance will be given to the effects of environment.

150. What are acquired characters? — The use of the term acquired characters to indicate entirely different facts has given rise to some confusion in the consideration of this subject. In one sense no character is ever transmitted. Only the determiners which give direction to the developing characters of the animal or plant are actually inherited. The characters themselves develop out of and are determined by the fundamental constituents of the germ substance. Thus the appearance of the secondary sexual characters of the male at puberty are not acquired characters in the biological sense. Acquired characters are modifications of the somatic cells which are induced by environment, use or disuse, accidents or any other influences acting upon the body-cells in such a way as to change their normal development. According to Weismann, if the modifications are the result of the presence of determiners in the germ-plasm, then they are not properly designated as acquired characters.

151. Somatoplasm and germ-plasm. — From the viewpoint of the biologist, all organic beings which reproduce sexually are differentiated into two very clearly defined groups of cells: the somatic group, which includes all the cells concerned in the processes of nutrition, including digestion, absorption and assimilation; the nerve cells and all other cells involved in the physiological activities of the organized being, except the reproductive cells.

Distinct and apart from the soma- or body-cells are the germ-cells, the function of which is to provide for the reproduction of the species. Weismann was the first to point out clearly the very sharp division between the fundamental organization and function of these two groups. The somatoplasm has its origin in the germ-plasm, and the direction of its development is controlled by the determiners in the germ-plasm, but the germ-plasm is not influenced in any fundamental way by the somatoplasm. The soma is to be regarded in the nature of a habitat for the successful activities of the germ-cells. In a sense, the somatoplasm has no more influence upon the developing germ-cells harbored within the soma than does the soil upon the trees which may develop from the seeds planted on its surface.

The germ-plasm is continuous. It is a part of the

germ-plasm of the preceding generation. If this distinction and differentiation is kept clearly in mind, it will help materially in the discussion of the inheritance of acquired characters. If we accept Weismann's definition of acquired characters, that they are somatic modifications which do not have their origin in the germplasm, we have to discuss only such somatic modifications as may be acquired by the animal during its lifetime.

152. Examples of acquired characters. — Among the common causes of the changes in the soma must be grouped environment, including food and climate; use and disuse of parts; disease and accidents. Each of these acting separately or all of them acting together may cause profound changes in the external form and development of plants and animals.

153. Food supply. — Perhaps no other single environmental influence is responsible for such profound changes in the external form of plants and animals as the food supply. The Kerry cattle of Ireland are a diminutive race of cattle which have been long subjected to conditions of scant supply of innutritious food. Their size and hardy character must be regarded as a more or less successful adaptation to their environment. When young calves of the Kerry breed are surrounded with conditions in which they are supplied with a generous and nutritious food, they increase in size and come to maturity at an earlier age.[1]

The Shetland pony in the barren islands of Shetland, gathering a scant subsistence from the inferior grasses and forage plants, develops into one of the most diminutive races of horses known to man. The same race of horses transplanted to the fertile regions of Great Britain

[1] Miles, "Principles of Stock Breeding," p. 100.

or America increases in size as the result of a better food supply.

Woltereck, by changing the food supply only, in hyalodaphnia, succeeded in changing the percentage of the height of the head to that of the body from 40 to over 90.

The remarkable influence of the amount of food supplied to an animal during its development in modifying the somatic cells and changing the external form of animals, is well illustrated by the unpublished results of an experiment conducted at the Missouri Experiment Station.[1] In this investigation, the animals were maintained for long periods upon different planes of nutrition.

For the purpose of this investigation the animals were divided into three groups. Group one (see Plates VII, VIII and IX) was supplied with a generous ration calculated to furnish to the animal all the nutrients it could utilize, and produce the maximum growth and development, including the laying on of fat. The treatment given to group one resulted not only in very rapid growth and development, but as the excessive feeding was long continued, the animals laid on unusual and excessive amounts of fat.

The ration supplied to group two (see Plate VIII, lower) was intended to provide such a quantity of food as was sufficient to produce strong, healthy growth and development, but insufficient for laying on any considerable amount of fat. This ration resulted in normal, healthy growth, but the food supplied was constantly below the desires and appetites of the animals. There was no time during the experiment at which the animals would not have consumed more food.

[1] Waters and Trowbridge, Unpublished Data from the Missouri Experiment Station.

Group three (Plate VII, lower) was limited markedly in the amount of food which the animals were permitted to consume. The scant ration given to group three did not prevent the animals from growing, but it did prevent the animals from making a normal growth, and prevented the normal deposition of fat within the body tissues, which seems to be a favorable factor in inducing healthy growth and development.

The illustrations represent typical animals in the three groups. These illustrations give a general impression of the changes in body form which are readily apparent to the eye, resulting from the methods of treatment of these various groups.

154. Influence of the amount of food on body weight. — A record of the changes in the body weight of an animal is not the most accurate measure of the influence of any environmental factor, but it is sometimes very useful, and often the best measure available. In this experiment, animal 501, which was fed for forty-seven months on a full ration, attained in that time a total weight of 1965 pounds. Animal 512, fed a medium ration for forty-eight months, attained a weight during that period of 1224 pounds. Animal 500, belonging to the low-fed group, weighed, at the end of the forty-eight months' feeding, 1042 pounds. The differences observed in these animals must be due entirely to the differences in the amount of food supplied, as they were subjected to identical conditions in all other respects. The experiment would have been still more valuable if the animals in the three groups had been fed on the different rations from birth. As a matter of fact, the animals in all the groups were generously fed during the first five months of their lives.

155. Food supply and body changes. — All the evidence available points to the fact that the domestic animals have inherited not only a tendency to reach a certain size and form, but that they have inherited a strong tendency to attain a given size at a certain age. Thus, in these investigations at the Missouri Experiment Station, there is evidence to show that the animal organism makes desperate efforts to grow, even when the food supply is greatly limited. In one case a calf nine-and-one-half months old was fed a limited ration which resulted in a loss of weight amounting to 82 pounds in six months. During the same period this animal increased 8.1 per cent in height and 14.1 per cent in length of head.[1]

156. Influence of limited food supply from birth. — The amount of food supplied to a growing animal in a large measure determines not only its ultimate development, but the rate at which the animals grow. In another investigation at the Missouri Experiment Station,[2] three animals were fed a full ration, a medium ration and a scant ration, respectively. Animal number 527 (Plate VII, upper) given a full ration, increased in weight rapidly, and at the end of 789 days weighed 1453 pounds. Animal 559 (Plate VIII, lower), given a medium ration intended to produce a normal growth, but not a fattening ration, at the end of 780 days weighed 813 pounds. Animal 551 (Plate VII, lower), given a scant ration, weighed at 777 days 512 pounds. It is, of course, possible that in these experiments the differences in body weight may be due to the

---

[1] Waters, Proceedings of the Society for the Promotion of Agricultural Science, 1908.

[2] Waters and P. F. Trowbridge, Unpublished Data of the Missouri Experiment Station.

amount of fat deposited in the tissues rather than to differences in the development of the skeleton and muscular tissues. The very great differences in weight noted are of the greatest significance if they represent differences in the fundamental skeletal and muscular tissues of the body. If they represent alone differences in the amount of fat, they are not so significant. The height of each animal as observed seems to indicate clearly that not only is the amount of fat deposited in the tissues directly determined by the amount of food available, but the skeletal growth also is profoundly influenced by the food supply.

There can be no question but that limiting the amount of food supplied to young animals has a profound influence upon the rate of growth as well as its ultimate size. This influence is to be found in smaller skeleton, probably arrested development of the muscular tissues, and a much smaller percentage of fat.

157. Telegony. — There was a time when eminent biologists and practical breeders firmly believed that the influence of the sire was not confined to his immediate offspring but the mother herself was in some manner so impressed with the characters of the sire that her subsequent progeny sired by entirely different males might take on the characters of the former sire. "The act of fecundation is not an act which is limited in its effect," says Agassiz,[1] "but that it is an act which affects the whole system, the sexual system especially; and in the sexual system the ovary to be impregnated hereafter is so modified by the first act that later impregnations do not efface that first impression." Darwin supported his belief in telegony by citing many cases among plants

[1] Massachusetts State Board of Agriculture, 1863, p. 56.

of "direct action of the male element on the mother form." He remarks further that "the male element not only affects, in accordance with its proper function, the germ, but the surrounding tissues of the mother plant."

Spencer likewise admitted the probability of the passage of the germ-plasm from the growing embryo into the maternal tissues and thus to the germ-cells. Weismann held that if any such influence in reality existed, it could be explained only on the theory that some of the sperm-cells of the male penetrated to the undeveloped ova and there accomplished a partial impregnation. Some practical breeders, of horses and dogs particularly, were so thoroughly impressed with the possibility of such an influence that they would not buy a highly bred animal that had borne offspring to another breed, believing that such a female could not be trusted to breed true. Some farmers in the mule-breeding districts have reported that horse foals from mares which had previously produced mules, sometimes possessed "mulish characters." These characters which are commonly possessed by the hybrid appearing thus in pure horse foals were supposed to have come about through the influence of the jack on the mother at some previous mating.

158. The Lord Morton mare. — One of the most striking cases of supposed infection was reported to the Royal Society [1] by Lord Morton in 1820.

The essential facts are discussed by Darwin.[2] In the year 1815 Lord Morton bred a seven-eighths Arabian mare of chestnut color to a quagga. The resulting

[1] Philosophical Transactions, 1821, p. 21.
[2] Darwin, "Animals and Plants under Domestication," vol. I.

offspring was a true hybrid having the same color and in other characters resembling the sire. Later in 1817, 1818 and 1821 the same mare was bred to a black Arabian stallion and from each mating produced a healthy foal which in every case was marked with stripes like the quagga sire, and resembling the sire also in the character of the mane. Lord Morton, in describing the particular resemblances of these foals, says: " Both in their color and in the hair of their manes they have a striking resemblance to the quagga. Their color is very marked, more or less like the quagga in a darker tint. Both are distinguished by the dark line along the ridge of the back, the dark stripes across the forehead and the dark bars across the back part of the legs." [1] The existence of stripes or bars on young foals is not uncommon, and their presence on the offspring of the Lord Morton mare may be explained on other grounds than the assumption that they were caused by the influence of a previous impregnation of the Arabian mare by the quagga sire.

159. The Penycuik experiments. — The possibility of tracing to some other source the real or fancied resemblances of the Arabian foals to the quagga sire of a previous mating, led Cossar Ewart to make a thorough investigation of the so-called infection theory. At Penycuik in 1895 Ewart repeated as closely as possible the breeding experiment of Lord Morton. In these experiments thirteen mares of varied colors and breeds produced a total of sixteen foals to a Burchell zebra. The same mares later produced twenty-two foals by Arab, Thoroughbred and Highland stallions. Ewart, in describing a typical

[1] Ewart, Bureau of Animal Industry Report, 1910, U. S. Dept. of Agr.

result, says:[1] "The first hybrid was born August 12, 1896, the dam being Mulatto, a black Highland pony lent by Lord Arthur Cecil."

"In 1897 Mulatto had a foal to Benazrek, a gray Arab stallion. As this subsequent foal of Mulatto was indistinctly striped, I was at first inclined to believe she had been infected by her first sire, the zebra Matopo; but when more richly striped pure-bred foals were obtained later by Benazrek out of Highland mares which had never even seen a zebra, it became evident that Mulatto afforded no evidence in support of the infection doctrine.

"Lord Morton's quagga counted for so little in the hybrid out of the chestnut Arab mare that its right to be regarded as a hybrid has been questioned. Matopo, however, proved so impressive that all his hybrid offspring plainly indicated their descent from a richly striped zebra.

"On the other hand, the subsequent foals (by Arab and other stallions out of the mares which proved fertile with Matopo) differed so profoundly from hybrids (even when, as was sometimes the case, they had bars on the legs and faint stripes across the withers) that they afforded no evidence that the first male influences 'the progeny subsequently borne by the mother to other males.'"

Baron de Parana in Brazil produced many zebra hybrids also from a true Burchell zebra, later rearing pure horse foals from mares that had previous zebra hybrids. In no single case did the stripes which did sometimes appear resemble closely the striping of the zebra.

A somewhat extended inquiry by Romanes in 1893

[1] Ewart, "The Principles of Breeding and the Origin of Domesticated Breeds of Animals," Report of the Bureau of Animal Industry, U. S. Dept. of Agr., 1910.

in British and American live-stock journals developed the fact that telegony is not as generally believed among breeders of the present day as has been generally reported. Sir Everett Millais, as the result of over fifty experiments with mammals and birds, found no conclusive evidence of infection.

Evidences of telegony, if it exists at all, ought to be easily collected in those regions where the production of mules is common. Darwin says,[1] " It is worthy of note that farmers in South Brazil (as I hear from Fritz Müller) and at the Cape of Good Hope (as I have heard from two trustworthy correspondents) are convinced that mares which have once borne mules when subsequently put to horses are extremely liable to produce colts striped like a mule."

These conclusions are not supported by Baron de Parana,[1] who reports, " I have many relatives and friends who have large establishments for the rearing of mules where they obtain 400 to 1000 mules in a year. In all these establishments, after two or three crossings of the mare and ass, the breeders cause the mare to be put to a horse because they believe that unless the mares are changed after producing three mules they become sterile. In all these establishments a pure-bred foal has never been produced resembling either an ass or a mule."

160. Telegony and mule hybrids. — In the horse-breeding districts of Missouri, large numbers of mules are produced annually. Many of the mares which have produced one or more mules are later bred to stallions and thus become the dams of horse foals. The jack and the stallion differ so widely in many important particulars that any marked tendency of horse foals to

[1] Report of the Bureau of Animal Industry, 1910, p. 123.

resemble a previous jack sire would be quickly observed. The writer [1] and C. B. Hutchison made an investigation of a large number of the horse offspring of mares which had previously foaled mule progeny from jack sires. Many of the horse foals examined were from mares that had produced more than one mule. In one case a mare had given birth to thirteen mule foals in succession and had then produced a horse foal. Many similar cases gave opportunity to observe a number of examples in connection with which full and favorable opportunity was present for the display of the influence of a previous impregnation. If, as some believe, the influence of a previous impregnation is cumulative and the dam becomes more and more completely infected by successive matings, these results should give an excellent opportunity to obtain some evidence on the theory of " infection " or " saturation."

A total of 168 mares were located that had given birth to mule foals and later had produced horse foals. Of this number 108 produced their first foal to a jack and later gave birth to horse foals. Among the number were forty mares that produced their first foals to a stallion, later producing mule foals and then again horse foals. The remainder were bred in a somewhat irregular manner, but all were alike in having produced horse foals following mule foals. The number of mares producing one or more mule foals each followed in every case by horse foals was as follows: Eighteen males produced one mule foal each, followed by a horse foal; twenty-two mares produced two mule foals each, followed by a horse foal; twelve mares produced three mule foals each, followed by a

[1] Mumford and Hutchison, Unpublished Data of the Missouri Experiment Station.

horse foal; twenty mares gave birth to four mule foals each, followed by a horse foal; eight mares gave birth to five foals each, and later to horse foals; ten mares observed produced six mule foals each, and afterward gave birth to horse foals; seven mares produced seven mule foals followed by horse foals; five mares produced horse foals after having foaled eight mules each, two mares dropped ten mules each, followed by horse foals; while one mare produced ten mule foals and then a horse foal and one mare was bred to a stallion and produced a healthy horse foal after having given birth to thirteen mule foals in succession.

161. Example of horse foals. — The horse foals from these mares were carefully examined and measured for the purpose of discovering any possible resemblances to the previous jack sire from which mules had been produced. The chief external characters which distinguish the mule from the horse are the size and form of the ears, head, feet, legs, body, mane, tail, disposition and voice.

The illustrations of dams and offspring give a fairly adequate idea of the generally uniform nature of the results which were found throughout the investigation.

In Plate X, lower, is shown the yearling offspring of a mare (Plate X, upper) that had previously produced seven mule foals and then gave birth to the animal shown in the illustration. A careful examination of the characters in which mules and horses differ showed that the offspring in this case had small, rather short, smooth head, small, short, pointed ears, a broad flat foot and rather broad hips and loins, with a well-rounded body. None of the characters of this yearling colt suggested in the slightest degree any evidence that it had been influenced by the fact of its dam having previously produced seven mule foals.

"Sallie" (Plate XII, upper) was foaled by a saddle bred mare fifteen years old and was the ninth foal from this mare. The previous eight foals were all mules. The mare "Sallie" was characterized by a small refined head of good quality, a small, short and pointed ear, rather scanty mane and tail, fine bone, broad and rounded foot and quiet and gentle disposition.

The mare "Hallie" shown in Plate XI was sired by a Standardbred stallion from the dam "Maude" (Plate XI, lower). Maude had previously given birth to ten female mule foals. At two years of age "Hallie" weighed 800 pounds, was fifteen hands high, had a short, small, refined head, short pointed ears, heavy and long mane and tail, broad flat foot and a rounded body. All of the external characters of this mare were clearly horse characters and not mule characters. There was no suggestion of any resemblance to the mule in any of its characters.

"Crewdson" (Plate XIII, upper), three years old, was sired by a Hackney stallion from the dam "Kate" (Plate XIII, lower), the latter having given birth to eleven mule foals followed by the horse foal "Crewdson." This animal was sixteen and one-half hands high and weighed 1000 pounds. The head was small and narrow, ear short and small, mane and tail medium heavy, foot broad and flat. "Crewdson" was characterized by a heavy mane and tail, small ear, refined head and broad rounded foot.

The cases cited and the illustrations used were selected because it was assumed that if the influence of a previous impregnation was likely to be exhibited at all it would appear in those dams which had produced a large number of mules followed by horse foals. No such evidence could be discovered. The 168 horse offspring from mares which had previously produced from one to thirteen mule foals each, gave no visible evidence of the existence of telegony. The external characters of the mule hybrid and the horse differ so widely in many important particulars that even a slight influence which might come through a previous sire should have been measurable.

It is true that the evidence is all negative, but it is nevertheless valuable because of the peculiarly favorable opportunity for the influence of telegony to assert itself. It is also interesting to note in this connection that a belief in the possible influence of a previous impregnation is by no means universal among practical breeders, if we may judge from the statements of farmers in the mule-breeding districts of Missouri. Very few breeders believed in the existence of telegony. Those who admitted their belief in the possibility of infection were unable to cite authentic instances.

162. Possibility of influence from a previous impregnation. — If the influence of the male is not confined to his immediate offspring but is extended to the mother in such a way that other progeny by other males may display some of the characters of the former male, then such influence must come about in one of two ways. The body (soma) of the mother herself may be so fundamentally changed by acquiring the characters of the male that she transmits such influence to her succeeding offspring sired by other males. The spermatozoa of the male may not only fertilize the fully mature and thus susceptible ovum but may travel through the generative organs of the female and eventually reach the ovaries where the developing and immature eggs might be so influenced that at a later time the fully fertilized egg would exhibit the results of fertilization by spermatozoa from two different males. The possibility of such double fertilization is extremely remote. The known facts regarding the successful fertilization of the egg are all against such an hypothesis. The egg is probably not susceptible to fertilization by the sperm except during a comparatively brief period which is coextensive with the heat period in the domestic animals. The immature ova still structurally a part of the ovary are not in proper condition to be fertilized. The probabilities are all against any such influence from this source.

Is it possible that the characters of the male may become impressed upon the pregnant female through the influence of the fœtus? If the body (soma) of the female is influenced in this way is this influence of such a nature that it can be impressed upon the embryo in the uterus? If it can, then the characters of a previous male may affect the later offspring by other males. Here again we must admit that the period of gestation may change the body of the mother to some extent, but it is extremely improbable that such change influences the offspring in any hereditary sense. As Rabaud[1] says, "Gestation naturally produces in the female a modification which we must suppose to be to some extent permanent. As a consequence, the female which produces a second offspring is no longer the female that produced the first offspring; whether the two gestations be due to the same male or to two different males, the fœtus of the second gestation evolves in conditions different from those surrounding the fœtus of the first gestation. But it does

[1] Étienne Rabaud, "Telegony," *The Journal of Heredity*, vol. 5, p. 389.

not undergo in any way the influence of the first male; in reality, what takes place is as if two different females were involved, mated with the same male or with two different males."

The conclusion seems very plain that the practical breeder has little interest in the subject of telegony. While it is difficult or impossible to prove by direct experiment that telegony does not exist, it is also true that no one by direct experiment has ever been able to produce any result which could not be explained on some other basis than that of telegony. All of the supposed cases of telegony can likewise be explained on some other basis than the assumption that a previous impregnation is lasting in its effect and may influence subsequent offspring.

163. Xenia in animals. — In plants the effect of cross-pollination in certain cases is to be observed in the fruits. Gardeners have long believed that the watermelons and citrons should not be planted in near-by locations because the pollen from the citron would injure the quality of the melon. Such injury really does not occur in this particular case but similar effects are present in other plants. When ordinary white corn is fertilized with pollen from a black variety, the grains so pollinated are black or mottled while other grains on the same ear are white. An explanation for this phenomenon is to be found in the fact that there are two cell nuclei in the ovum and two nuclei in the pollen cell. The primary nuclei of the two cells unite to form the daughter nucleus of the new cell. The two secondary nuclei likewise unite in the formation of the endosperm. In the case of cross-pollination of black and white corn it is the color of the endosperm which exhibits the influence of the crossing.

Among mammals the highly imaginative idea has been suggested that the influence of the male on the female (infection) may in turn be passed on to a later male mated with the infected female. A highly successful breeder of Shorthorn cattle in England pointed out to the author a pure-bred white cow with red ears from a registered sire and dam. This cow was marked like the wild white cattle of Chillingham Park, and her owner ascribed her markings to the fact that her dam had once dropped a calf from a Chillingham bull. The cause of the peculiar markings of the cow in this case could not have been derived in the manner described. The case is cited here only as an example of the highly improbable notion that there is some relation between the phenomena of telegony and xenia. Xenia among mammals is unknown.

164. Xenia among poultry. — Many breeders of poultry have believed that the eggs of the domestic fowl may be influenced in size, form and color by the male bird. Observations by Nathusius and later by Holdefleiss [1] gave evidence of paternal influence on the color of the eggshell. Holdefleiss mated Plymouth Rock hens with a Leghorn cock. The Plymouth Rock uniformly lays brown eggs while the Leghorn lays a pure white egg. The eggs deposited by the Plymouth Rock hens from this mating varied in color from dark brown to white. The evidence seemed so clear to the investigator that he was led to conclude that "The color of eggshells shows the influence of the paternal strain; there is therefore evidence of xenia." More recently Walther [2] of Giessen after a series of careful experiments has reported results which

---

[1] Holdefleiss, "Berichte aus dem biologische Lab.," Univ. Halle, 1911.

[2] Walther, "Landwirtschaftliche Jahrbücher," 1914.

do not confirm those of Nathusius and Holdefleiss. Walther's investigations included not only color but also size, form and glossiness. His conclusions were that the paternal parent has no influence on the size, shape and glossiness of the eggs. His results on the color of eggs are not conclusive but tend to discredit the theory of xenia in fowls. From all the evidence available, it would seem that the possibility of xenia in fowls is not satisfactorily determined, and further investigation is needed to settle this supposed influence of the male bird on the color of eggs.

OBJECTIONS TO THE THEORY THAT ACQUIRED CHARACTERS ARE TRANSMITTED

The trend of opinion of modern biologists has been further and further away from the belief in the possibility of the inheritance of acquired characters. As our knowledge of cellular biology has increased and we have been able to study the mechanical processes which are concerned in reproduction and heredity, it has become more and more apparent that Weismann's view of the essential separateness of the soma-cells and germ-cells is substantially correct.

165. No mechanism for the inheritance of acquired characters. — There is no mechanism by means of which somatic modifications may impress themselves fundamentally upon the germ-plasm. In fact, the development of the soma-cells is made possible by determiners in the germ-cell. The very fact that the soma-cells have been able to respond to external influences and develop in a direction somewhat different from the average of the species is sufficient evidence that the determiners which were responsible for the tendency of the organism to develop in the given direction were already present in the germ-plasm. As Walter aptly remarked, "Not only the development of the race which we call evolution, but also the determination of the individual in heredity, is a chain of onward-moving sequences like the succession of events in history. It is hard to see how recent events can influence preceding events. It is hard to see how the water that has gone over the dam can return and affect the flow of the river upstream in any direct way. It is likewise hard to see how differentiated somatoplasm, which represents the end stage of a successive series of modifications, can make any definite impress upon the original germplasmal sources from which it arose." [1]

Even Darwin found difficulty in believing in the inheritance of acquired characters. His theory of pangenesis which assumed that each somatic cell added to the circulation a minute granule which later found its way to the germ-plasm is not substantiated by later investigations.

The evidence presented to prove that somatic modifications are actually transmitted from parent to offspring is not conclusive. It is certain that acquired characters cannot be transmitted unless the germ-plasm has been definitely changed by reason of somatic influence. The evidence of such influence upon the germ-cell is entirely negative. Definite experiments carefully planned for testing the possible effect of such influence have been inconclusive.

166. The inheritance of disease. — The earlier writings on animal breeding contain numerous references to the possibility of inheritance of disease. Many examples

[1] Walter, "Genetics," 1913, p. 85.

are recorded among the domestic animals of supposed cases of the heredity of pathological conditions. Among the diseases which have been regarded as hereditary are tuberculosis, melanosis, broken wind, specific ophthalmia, blindness, spavin, ringbone, curb and many other diseases. The discussion of the transmission of diseased conditions of the organism brings forward again the entire question of the possible inheritance of acquired characters. In general it may be said that recent researches in biology have resulted in demonstrating that many diseases which were formerly regarded as transmissible are no longer believed to be transmitted through inheritance. This certainly applies to all diseases which are contracted after birth. Some diseases which are the result of a definite variation in the germ-plasm will of course be transmitted. We must therefore clearly distinguish between inborn disease and acquired disease. Certain diseases or defects are undoubtedly transmitted from parent to offspring. Whenever such defects represent changes in the germ-plasm, then such defects will be as certainly transmitted as any other character of the animal.

Such defects which may be transmitted are deafness, color-blindness, idiocy and possibly rheumatism, gout and insanity. In the latter diseases their apparent transmissibility may be the result only of a predisposition.

167. Acquired diseases. — Many diseases of the domestic animals are acquired after birth. A large number of pathological diseases are due to infection. All such diseases are no more certainly transmitted than are other acquired characters. Tuberculosis is the result of infection by a specific germ and this germ may be acquired under certain conditions by the animal organism. In the case of bone diseases of horses, which were for a long time held to be inherited, it is probable that the development of such unsoundness is due largely to the action of external factors. In other words, they are acquired. It is true, however, that certain individual animals or families are much more subject to bone disease than other families. In such case we must recognize a predisposition to disease.

168. Congenital disease. — The fact that a disease exists at birth is not always adequate evidence that disease has been inherited. It is possible for certain germ diseases to infect the fœtus in utero. It is also probable that the ova or spermatozoa may under certain conditions carry the infection and this infection may be present and active during the prenatal life of the animal.

169. Predisposition to disease. — Many diseases appear to " run in families." For this reason we have recognized the fact that the members of certain families are subject to tuberculosis, gout, rheumatism, imbecility, insanity or other diseases. In all these cases there exists a predisposition on the part of the members of a given family to acquire the diseased condition. Through weakness of certain organs or general lack of constitutional vigor, the infective germs of many diseases may overcome the resistance of the animal organism to disease. This predisposition is certainly and often strongly inherited. From the standpoint of the practical animal breeder, therefore, a predisposition to disease may be quite as significant as the actual transmission of the disease.

170. Immunity. — An interesting correlated fact is the probable inheritance of immunity from certain diseases. Certain families seem to possess immunity from certain diseases, such as smallpox or diphtheria. It is not possible at this time to enter into a discussion as to the nature of immunity, but undoubtedly the known immunity of certain individuals or families to certain diseases may become the basis of important future improvements in the domestic animals. In recent years it has been found that certain hogs are immune from hog cholera. Experiments have been suggested to determine whether or not it would be possible to develop a race of swine immune to this dreaded disease.

It is possible for animals to acquire immunity through vaccination of actual infection of the disease itself. Such immunity is not transmitted. It is doubtful whether there exists any congenital immunity, but investigations along this line might be fruitful of results.

CHAPTER IX

HEREDITY AND SEX

THE chief function of both plants and animals is to live and reproduce. In many wild forms the powers of reproduction are little short of marvelous. A single plant of purslane may produce a million seeds. Man is less productive than most other mammals, but masses of population have been known to double in twenty-five years. At this rate in 1000 years there would not be standing room on the earth for his children.

The natural increase of plants and animals is not realized because of unfavorable conditions. The number of animals that can exist on a given area is limited. If too many are born, some must inevitably die. Others are destroyed by enemies, while still others are poisoned by substances which accumulate within their own bodies.

The kinds of reproduction have already been mentioned under the general subject. The simplest form of reproduction is by cell division. This method of reproduction is chiefly found in unicellular organisms like the amoeba and paramoecium. The most common method of reproduction is by eggs. Egg production is almost universal among both plants and animals. But the egg generally is inert and incapable of development into a new individual until it has been fertilized. Herein lies the apparent reason for differentiation into male and female sexes.

171. The significance of conjugation and fertilization. — The real purpose of fertilization is not well understood. Biologists are not yet able to speak with positive assurance as to the real character of the actual biological phenomena which result from conjugation of the male and female germ-cells. Bütschli believed that fertilization was a process of rejuvenation. This idea involves the assumption that the union of the germ-cells of two weak individuals will result in the production of a strong individual.

Jennings, in a series of very skillful and carefully controlled experiments with paramœcium, found that after conjugation the rate of division was not accelerated but was actually slower. Paramœcium which had been artificially weakened and their rate of division retarded when allowed to conjugate was not in most cases benefited. Some were apparently benefited, but in all cases the rate of division was slower than in cultures in which the paramœcium was well nourished. In other words, conjugation was of advantage to some and not to others. Jennings concluded that conjugation is for the purpose of bringing about a recombination of characters. Some of them are very beneficial and will persist and multiply, others are disadvantageous and these will fail to live and reproduce. The combinations most likely to persist are heterozygous.

As Morgan[1] has explained, " The meaning of conjugation, and by implication, the meaning of fertilization in the higher forms is from this point of view as follows: In many forms the race, as a whole, is best maintained by adapting itself to a widely varied environment. A heterozygous or hybrid constitution makes this possible,

[1] Morgan, "Heredity and Sex," p. 12.

and is more likely to perpetuate itself in the long run than a homozygous race that is from the nature of the case suited to a more limited range of external conditions." Whatever may be the real nature and purpose of fertilization, it is certainly true, as Wilson[1] remarks, that "the paternal germ-cell is the carrier of something which incites the egg to development, and thus constitutes the fertilizing element in the narrower sense."

172. Secondary sexual characters. — The sexes in the higher animals are differentiated, not alone by the possession of radically different essential organs of reproduction, but also by the possession of so-called secondary sexual characters. The more brilliant plumage of the male bird, the horns of the ram, and the greater development of the head and horns of the bull are examples of secondary sexual characters. Darwin regarded the secondary sexual characters as of great significance in sexual selection. As a result of sexual selection he believed that "generally, the most vigorous males, those which are best fitted for their places in nature, will leave most progeny."[2]

173. Secondary sexual characters and vigor. — Breeders of the domestic animals have long regarded the degree of development of the secondary sexual characters as an index of sexual vigor. Many have held that a male with the secondary sexual characters strongly developed was not only prepotent in the transmission of purely sexual characters but also in other characters which are desirable to man. Direct evidence is not available to show that because an animal is strongly developed in the secondary sexual characters, he is therefore prepotent

[1] Wilson, "The Cell," p. 230 (1911).
[2] Morgan, "Heredity and Sex," p. 101.

in the transmission of all characters. Sexual vigor is associated with the development of the secondary sexual characters, and sexual vigor is a desirable character in the domestic animals. The general efficiency of the reproductive process is undoubtedly correlated with the secondary sexual characters. The breeder, therefore, is making no mistake in emphasizing the importance of evidences of strong sexuality as indicated by the development of the secondary sexual characters.

174. Effects of castration and ovariotomy on the secondary sexual characters. — There is ample evidence of the close correlation existing between the essential organs of sexual reproduction and the secondary sexual characters. The full development of the secondary sexual characters is closely connected with sexual maturity. In the Merino breed of sheep, the males are always horned while the females are hornless. If the male is castrated before the horns begin to develop, the horns fail to grow and the wether remains hornless. If the males are castrated after the horns have started to develop, the horns cease to grow. Marshall, in experiments with Herdwick sheep, a breed in which the males are supplied with large coiled horns and the females are hornless, found that castration at varying ages invariably caused a cessation in the growth of the horns of the male. When the ovaries of the female were removed, there was no apparent tendency toward the growth of horns, although small scurs appeared in one spayed ewe that was kept for seventeen months after removal of the ovaries. Marshall concludes, "The development of horns in the males of a breed of sheep in which well-marked secondary sexual differentiation occurs (as manifested especially by presence or absence of horns) depends upon a stimulus arising in the testes, and this stimulus is essential, not merely for the initiation of the horn growth, but for its continuance, the horns ceasing to grow whenever the testes are removed."

"The removal of the ovaries from young ewes belonging to such a breed does not lead to the development of definitely male characters, except possibly in an extremely minor degree."[1]

Arkell[2] crossed Merino ewes with a Southdown ram (hornless). The sons of this cross had horns. The factor for horns in this case must have been present in the Merino mother, herself hornless, but the full development of horns cannot take place except the male glands are present and functional.

175. Effect of transplanting sexual glands. — The investigations already described seem clearly to point to the conclusion that some stimulus to the development of the secondary sexual glands exists in the testes and ovaries.

Steinach transplanted ovarian tissue from a female guinea pig to the tissues of a castrated male. The result was to cause the rudimentary mammary glands of the male greatly to enlarge and the male came to resemble the female in certain characters.[3]

A remarkable experiment is described by Goodale[4] in which the ovaries of a female Mallard duck were entirely removed and the plumage became like that of the male Mallard.

176. Effect of internal secretion. — The secretions of various internal organs have a profound influence upon the development of the individual. These secre-

[1] Marshall, Proceedings Royal Society (London), ser. B, 85, 1912.
[2] Arkell, New Hampshire Agr. Exp. Sta., Bul. 160.
[3] Morgan, "Heredity and Sex," 1913, p. 140.
[4] Goodale, *Journal Experimental Zoölogy*, 10.

tions, called "hormones," emanate from various glands and perhaps from most of the internal organs. If the thyroid and parathyroid bodies are removed from the body, death follows. The destruction of the pituitary glands in man causes the bones of the hands, feet and jaws to enlarge (gigantism), causing death.

It is probable that the milking function in the domestic animals has some connection with the activities of a specific "hormone" which is essential.

177. Sex-linked characters. — Certain characters are so closely related to sex that their transmission is influenced by such relation. These characters have been called sex-linked or sex-limited characters. They are to be distinguished from secondary sexual characters.

178. Color-blindness. — Men are much more frequently color-blind than women. Color-blind men do not transmit color-blindness directly to sons, but to grandsons through their daughters. The daughters of color-blind men are not themselves color-blind, but tend to transmit this deficiency to their sons. Color-blindness in the daughter could be produced only when the father was color-blind and the mother possessed the power to transmit color-blindness. Color-blindness is the result of some defect in the germ-cell. This factor which is associated with an x chromosome appears twice in the ovum and only once in the sperm. A similar condition is found in the pomace fly [1] (*Drosophila ampelophila*). This form has normally red eyes, but this apparently is a unit character, sex-linked in transmission. An interesting case of sex-linked heredity is found in the Barred Plymouth Rock fowl.[2] Pure barred fowls when mated produce

[1] Castle, "Heredity and Eugenics," p. 75.
[2] Castle, "Heredity," 1911, p. 170.

only barred offspring. When the male Barred Rock is bred with a non-barred variety, the offspring of both sexes are all barred. If the female Barred Rock is mated with non-barred breed, the offspring will be about one-half barred and the remainder non-barred. The barred offspring are always males, while all the females are non-barred. The barred character is, therefore, sex-limited. The Barred Rock female is heterozygous and the male homozygous. The pure Barred Rock breed transmits the barred quality because the male is pure in respect to the barred character. The same result follows if a cross-barred male is mated with barred females. The explanation of sex-linked inheritance is probably to be found in the existence of some plus element in the egg which is not found in the sperm.

179. Controlling the sex of offspring. — In many of the domestic animals, the sex of the individual determines its peculiar value and usefulness to mankind. If some method of breeding could be devised which would result in the production of the particular sex desired, it would be a great economic gain. That such attempts have been made by both ancient and modern breeders is made clear from an examination of the literature of the subject from the earliest times to the present. Because of the more or less general belief among practical breeders in the possibility of controlling sex, it seems necessary to consider briefly some of the more widely held theories of sex control.

180. Age or vigor of parents. — Two investigators, Sadler[1] (1830) in England and Hofaker (1823) in Germany, collected statistics representing more than 2000 births. Their statistics showed that when the father is

[1] Carpenter, "Human Physiology," p. 1015.

older the larger number of the offspring are males, and when the mother is older the children tend to be females. These results have been confirmed by Göhlert, Boulanger and Legoyt. Many practical breeders have also cited special cases in which the sex offspring seemed to follow that of the older parent. Giroude Buzareingues [1] reports results from breeding young immature ewes to strong mature rams. The proportion of sexes was 80 males to 35 females. When young rams were used as sires, the proportion of sexes was 53 males and 84 females.

SUMMARY OF STATISTICS BEARING ON RELATIVE NUMBER OF MALES AND FEMALES [2]

Observer	Number of Births	Locality	Father Older. Proportion of Males to 100 Females	Father Equal Age. Proportion of Males to 100 Females	Father Younger. Proportion of Males to 100 Females	Average Proportion of Males to 100 Females	Remarks
Hofaker	1,996	Tubingen	117.8	92.0	90.6	107.5	—
Sadler	2,068	England	121.4	94.8	86.5	114.7	—
Göhlert	4,584	England	108.0	93.2	82.6	105.3	—
Legoyt	52,311	Paris	104.49	102.14	97.5	102.97	—
Boulanger	6,006	Calais	109.98	107.92	101.63	107.9	—
Noirot	4,000	Dijon	99.7	—	116.0	103.5	—
Breslau	8,084	Zürich	103.9	103.1	117.6	186.6	—
Stieda	100,590	Alsace-Lorraine	105.03	—	108.39	106.27	Contradictory
Berner	267,946	Sweden	104.61	106.23	107.45	106.0	Contradictory

[1] *Quarterly Journal of Agriculture*, vol. 1, 1828, p. 63.
[2] Geddes and Thompson, "The Evolution of Sex," p. 35.

The differences are not large and the number of observations are entirely too few to justify any sweeping conclusions. Steida [1] and Berner found no relation between age and the sex of offspring. The evidence of the influence of age on the sex of offspring is too conflicting to be conclusive.

181. Comparative vigor or sexual superiority. — Various authorities have attempted to explain the proportion of sexes on the theory that the sex of the offspring will correspond to that of the more vigorous or " superior " parent. Darwin, Richarz, Hough and others regarded the male as to a certain extent a superior organization, and male offspring would result when the reproductive functions of the mother were particularly well developed. The evidence available is not sufficient to give this hypothesis any particular importance in practical breeding.

182. Nutrition and sex. — That the nutritive condition of the parents, particularly the mother, at the time of fertilization and before has a preponderating influence on the sex of offspring has been long believed. Yung found that under certain conditions regarded as normal, the proportion of sexes in tadpoles was about 57 females to 100 males. By feeding the tadpoles beef, fish and frog's flesh, the percentage of females enormously increased, being in one case 92 females to 8 males. An interesting case illustrating the connection of nutrition and sex is found in bees. The swarm of bees is composed of workers (imperfect females), drones (males) and the queen (a perfect female). The drones are hatched from unfertilized eggs. The queen and workers are developed from fertilized eggs, but perform a very different rôle in life. The queen becomes the mother of new generations,

[1] *Ibid.*

while the worker bees are sexually imperfect. It seems to be true that the eggs developing into worker bees and queens are identical. The one becomes a queen as the result of a "royal" diet, while the worker larvæ are fed on a "common" diet and develop into the non-fertile female. Investigations with wasps by Von Siebold [1] and butterflies and moths by Mrs. Treat suggest a real connection between nutrition and sex offspring. Schenk found that starvation produced fewer males, but later the same condition resulted in producing more males. Düsing reported that among the Swedish nobility the proportion of sexes was 98 boys to 100 girls, and in the Swedish clergy 108.6 boys to 100 girls. Punnett submits evidence that in London more girls are born among the poor than the rich. In most of the cases cited there are too many other factors involved to justify the conclusion that nutrition alone is responsible for the proportion of the sexes.

183. The maturity of the ovum. — The degree of development or maturity of the ovum itself at the time of fertilization has a controlling influence on sex in the opinion of some breeders. If fertilization occurs during the early part of the heat, the offspring will be female. If the ovum is fertilized later in the heat, the offspring will tend to be males. Thus Thury [2] of Geneva says, "The sex depends upon the degree of maturity of the egg at the moment of fecundation, that which has not reached a certain degree of maturity producing the female, and, if fecundated when this point of maturity has passed, producing a male." These results are not altogether consistent with ordinary farm practice or with the experi-

---

[1] Rolph, W. H., "Biologische Probleme," Leipsig, 1884.
[2] *Country Gentleman*, 1864, p. 12.

ments of others. In ordinary breeding operations on the range, the bull invariably runs with the cows and breeding occurs during the first part of the heat. If the sex is determined to any extent by the particular stage of heat, then in this case the offspring should be largely female. But such is not the case, the proportion of the sexes is practically equal, subject to seasonal or other variations not entirely explainable. Thury's results have not been satisfactorily confirmed. Miles [1] recorded the proportion of sexes among cattle and sheep on the Michigan Agricultural College farm for a period of ten years. The results were as follows: Sheep, 102.5 males to 100 females. Cattle, 118.4 males to 100 females. All the animals were bred during the first part of the heat and the offspring should have been more largely female. The proportion of the sexes seems to be subject to wide variation, and therefore any investigations of this kind must necessarily include large numbers of animals and the observation be carried over a series of years to make the results of any value.

184. Seasonal variations in proportion of sexes. — The proportion of the sexes is subject to wide variations apparently due to seasonal influences. Quoting again from Miles: [2]

"In 1864 and 1865 the bull-calves were 2.5 to 1 heifer; in 1866 and 1867 the heifers were considerably in excess; in 1868 and 1869 the heifers were nearly 2 to 1 bulls; in 1870 the bulls were decidedly more numerous; and in 1871 and 1872 there were more than 2 bulls to 1 heifer. In 1872 there were 2 rams to 1 ewe, and the bulls were nearly in the same proportion to the heifers, which would

[1] Miles, "Stock Breeding," 1878, p. 265.
[2] *Ibid.*, p. 299.

seem to indicate some peculiar influence of the season in favor of the males. In 1871, however, the bulls were largely in excess of the cow-calves, and there was quite as decided a preponderance of females among the sheep."

185. Sex cannot be controlled by external conditions. — The most recent investigations of sex determination lead to the belief that sex is predetermined in the germ-cell. It is not subject to change through any change in the environment, as nutrition or temperature. The germ-cell contains a determiner for sex as it contains determiners for other characters.

Castle concludes,[1] "If, as has been suggested, the determination of sex in general depends upon the inheritance of a Mendelian factor differentiating the sexes, it is highly improbable that the breeder will ever be able to control sex. Male and female zygotes should forever continue to be produced in approximate equality, and consistent inequality of male and female births could result only from greater mortality on the part of one sort of zygote than of the other."

The same idea is similarly expounded by Morgan.[2] "If these observations are confirmed, they show that in man, as in so many other animals, an internal mechanism exists by which sex is determined. It is futile, then, to search for environmental changes that might determine sex. At best the environment may slightly disturb the regular working out of the two possible combinations that give male or female. Such disturbances may affect the sex ratio but have nothing to do with sex-determination.

[1] Castle, "Heredity," 1911, p. 180.
[2] Morgan, "Heredity and Sex," p. 248.

CHAPTER X

VARIATION

No biological fact is more clearly recognized than the tendency of all plants and animals to vary. In a broad way, individuals resemble their parents. Through heredity the qualities of the parent are transmitted to the offspring. A horse begets other horses and not pigs. We do not gather grapes of thorns nor figs from thistles. Thus has come about the old adage " like begets like," but this aphorism applies only within certain limits and fails to take into account the inexorable law of variation by which the offspring is never exactly like the parent. It is true that the offspring of a horse will always be a horse and not a cow. It is even true that the progeny of a trotting horse will be a trotting horse, but the keen judge of horses is able to discern slight variations in form, color, disposition or ability to perform.

No two animals are exactly alike. The difference may be slight or very marked. From the same sire and dam, the offspring may differ widely in character among themselves. In a large family of boys, the physical characters and mental dispositions of the individual members of the family may be very different. It has happened in the history of trotting and running horses that own brothers have varied widely in their ability to win in speed competitions.

186. Importance of variability. — The inherent tendency to vary which exists in all organic beings makes the improvement of domestic races possible. Inorganic compounds are fixed in composition and physical character. Pure gold cannot be improved. The individual animal has within its own organic constitution not only the capacity for, but a noticeable tendency toward, variation from the characters of its ancestors. This fact lies at the very foundation of the successful improvement of domestic animals. If it were true that the offspring was identical in character with the parent, then improvement would be impossible.

In developing new varieties, the breeders' efforts are sometimes first directed toward encouraging the tendency to variation. If the particular form which the breeder is endeavoring to improve possesses an unusual tendency to vary, then a large number of new characters or new combinations of characters will occur. Some of these will be desirable, but many will be less desirable than in the parent forms. The individuals which possess the most useful and valuable characteristics will be preserved by selection. After desirable variations have occurred, the next step is to fix these by heredity so that they may become racial characteristics and be transmitted with some degree of regularity from parent to offspring.

187. Morphological variations. — The variations which are of interest to the animal-breeder exhibit many different forms.

Morphological variations are those affecting the form. These may be merely differences in size, as in the case of two pigs possessing the same characters and the same relative development of characters but differing in size.

Some individuals are dwarfs while others are giants in size. This variation in size is due to a difference in the number of cells rather than in the size of the cells. The increased number of cells found in the larger individuals represents excessive cell division, while the abnormally small are the result of imperfect and arrested cell division.[1]

The cause of undersized individuals among animals cannot always be determined. It is certain, however, that insufficient food or food which is deficient in the necessary elements of nutrition is a common cause of undersized animals.

It is also true that when females become pregnant while still growing and immature, their growth receives a sudden check. The arrested development is not so much due to the effects of pregnancy as to the strain of lactation. The growth of well-fed females is probably not to any appreciable extent checked by pregnancy, but the physiological requirements for the production of milk are severe and the development of immature mammalian females is abruptly arrested during the period of lactation.

Differences in size are of relatively less importance than morphological differences arising from variation in the relative development of the parts of the body. This may be illustrated by the variations in meat animals. Beef cattle may possess a broad back, well-sprung rib, and a thick covering of flesh over all parts, or they may lack these highly desirable qualities.

188. Physiological variations. — Changes in the functional activities of animals are frequent and important. The average domestic cow produces not more than 150 pounds of butter in a year, but selected herds may produce

[1] Davenport, "Principles of Breeding," p. 27.

500 pounds of butter in one year. Some individuals are very fertile, others may be sterile or markedly deficient in this very desirable quality.

189. Meristic variation. — All plants or animals develop in accordance with a certain symmetrical pattern. A quadruped has four legs, two ears, two eyes, and two sides of similar character. A deviation from the characteristic plan or pattern of the species is called a meristic variation. Such variations are of very great importance among plants, but are of little practical significance to the animal-breeder. Examples of meristic variation are to be seen in the doubling of flowers, the stooling of grains, and the production of four-leaved clovers. Among animals the growth of extra fingers and toes, and the development of an abnormal number of vertebræ or ribs, are not uncommon. An interesting example of this type of variation is found in the development of extra mammæ. Supernumerary nipples in mammals are a common form of variation among humans. Bruce[1] found fourteen cases of extra mammæ among 2311 females examined. Most male mammals are supplied with rudimentary nipples, and curiously there seems a greater amount of variation among males than females. The same authority quoted above found forty-seven cases of multiple nipples among 1645 males examined. In one case a woman[2] is reported to have possessed five pairs of nipples. The presence of a larger number of mammæ than the normal has been regarded by some as evidence of greater fertility.[3] A variation in the opposite direction resulting in the development of a smaller number of digits than the normal

---

[1] Bateson, "Materials for the Study of Variation."
[2] *Ibid.*, p. 183.
[3] Bell, "Multinippled Sheep."

is to be found in the case of the " mule-footed " hog. In this instance, the split hoof has united into one solid hoof like the horse, and this variation is strongly inherited. Meristic variations are often transmitted by heredity.

190. Functional variations. — In our discussion of variation up to this point, we have chiefly concerned ourselves with such changes as are exhibited in the form of animals. We have now to consider those important modifications which affect the performance of the individual. An animal may retain the same form as the average of the race but be vastly more efficient in the performance of some one or more of the physiological functions. This type of variation is one of the most important to the animal-breeder. The value of many of the domestic animals depends entirely upon their functional development. The chief advantage possessed by some domesticated animals over their wild progenitors is due to their higher functional efficiency. This is illustrated by the domestic cow, the improved breeds of wool sheep, and the trotting horse. Such differences in form between the wild and domestic sorts as are present are chiefly significant as indicating the correlation which exists between function and form. The changed form, if it exists, is incidental and the result of functional influence.

The valuable variation which the breeder has emphasized in his selection has been the ability of the animal to produce economically and largely some valued animal product. The breeder did not in the beginning consciously select variations in form. It is true, however, that the efficient performance of animals is to a certain extent correlated with the form, and it follows, therefore, that we may within certain rather narrow limits rely upon the external form as an indication of functional efficiency.

191. Examples of functional variation. — In most cases the functional variations in the domestic animals which have become valuable to man are modifications of natural functions possessed by all wild animals living under natural conditions. The present stage of development in domestic forms is due to artificial selection practiced by man. The variations which have been desirable have been preserved through selection and a gradual improvement in functional efficiency has resulted. These variations may arise through sudden mutations, but such marked changes in function are probably less common than similar mutations in form. Variations in the functional activities of animals are of great economic importance to the breeder of domestic animals. It is important to know with some degree of definiteness the extent of variation in function, as such knowledge will give some idea of the probable limits of improvement. The following examples of functional variation will be useful in helping to determine the limitations of improvement. Such examples may be greatly multiplied by search of the literature of the subject.

192. Variation in fertility of animals. — It is well known that there are wide differences among individual animals and among races or breeds in their ability to produce large numbers of offspring. The quality of fertility is one of great practical importance and is readily transmitted by heredity. Miles[1] has recorded the case of a cow belonging to a French farmer which produced nine calves at three births, four at the first, three at the second, and two at the third.[2] A Teeswater ewe belonging to Edward Eddison produced four lambs in 1772 when

---

[1] Miles, "Stock Breeding," p. 131.
[2] "British Husbandry," vol. II, p. 438.

two years old; "in 1773, five; in 1774, two; in 1775, five; in 1776, two; and in 1777, two. The first nine lambs were lambed within eleven months."[1] The Country Gentleman reports the case of a sow that produced twenty-three pigs at one birth. The above examples are of unusual cases of fertility and are so far above the average fertility of the respective races of animals that they represent a marked departure from the established type. Other things being equal, those races of domestic animals which are most fertile are most profitable. One of the qualities of economic importance which commends certain breeds to the practical farmer is the quality of fertility. This is especially true among swine and sheep.

193. Variation in the milking function. — The milking function in domesticated animals is of special interest for the reason that it is probably the most valuable single functional variation in the domestic animals. The improvement of the domestic cow in the direction of greater functional efficiency in the production of milk and butter has been little short of marvelous. This improvement has not only resulted in developing an animal with a capacity to produce enormous quantities of milk, but the efficiency of the cow in the economic utilization of food has been no less noteworthy. The highly improved domestic cow is able to utilize the raw products of the farm, consisting of grain, hay and grass, and produce from these a larger amount of human food than any other domestic animal. The high efficiency of the milk cow as compared with the beef steer is clearly shown by the records kept at the Missouri Experiment Station.[2] A

[1] Culley, "Live Stock," p. 123.
[2] Eckles, "Dairy Cattle," p. 6.

Holstein Friesian cow produced 18,405 pounds of milk in one year. This year's production from one cow contained 2218 pounds of dry matter. At the same station, the carcass of a 1250-pound fat steer was analyzed. The steer was twenty-one months old and had been fed generously from birth. The steer's carcass contained 548 pounds of dry matter, or a little less than one-fourth the amount of dry matter produced by the cow in one year. The dry matter recorded for the steer was for the entire carcass of the animal, including hide, horns, bones and intestines. The actual net weight of the dry matter of the edible portion of the fat animal was only 357 pounds. In other words, the Holstein Friesian cow produced six times as much edible human food in twelve months as was produced by the fattening and growing steer in twenty-one months. In the case just cited, it is true that the cow was far above the average in efficiency while the steer was a fair representative of a good average beef animal. The example is nevertheless a most excellent illustration of the very great development and improvement of the dairy cow in the direction of desirable variations. The foregoing records of the two animals did not include the dry matter consumed in the production of milk and beef. Such a record which would give the amount of dry matter required to produce a pound of dry matter in milk as compared with a pound of dry matter in beef would be of great interest. Such records have been kept at a number of American experiment stations.

The feed and milk record of Pedro's Ramaposa 181,160 has been given by Eckles,[1] and from this the dry matter in feed which is required to produce a pound of dry matter

[1] Eckles, Missouri Experiment Station, Research Bulletin, No. 2, p. 117.

VARIATION

in milk can easily be determined. The cow, Pedro's Ramaposa, during a period of one year produced 8522.9 pounds of milk which contained 1317 pounds of dry matter. In the production of this quantity of milk she consumed in feed 9362 pounds of dry matter. In other words, the consumption of 100 pounds of dry matter in the feed resulted in the production of 91.12 pounds of milk containing 14.06 pounds of dry matter. Stated in other terms, the cow here described produced one pound of dry matter in the milk for each 7.1 pounds of dry matter consumed in the feed.

194. Variations among different cows. — The milking function is hereditary and is a comparatively well-fixed character among the dairy breeds of cattle. It is true, however, that there still exist wide variations in the productive capacity of cows, even of the same breeding as well as those of different ancestry. The Illinois Experiment Station [1] in several tests has clearly demonstrated the wide differences which may exist between individuals.[2] Two native cows, Rose, nine years old, and Nora, six years old, were fed the same kind of a ration for twelve months. The amount fed was determined by the appetites of the animals. The table on the following page gives the essential facts of interest in this connection.

" Reduced to a like feed basis, for every 100 lb. of milk given by Nora, Rose gave 139.5 lb., and for every 100 lb. of butter-fat produced by Nora, Rose produced 180.7 lb."

Commenting on this test, Fraser says,[3] " As milk is

---

[1] Fraser, Illinois Experiment Station, Bulletins 51 and 66.
[2] Davenport, "Principles of Breeding," p. 78.
[3] *Loc. cit.*

nearly always valued by the amount of butter-fat which it contains, and Rose produced on the same feed basis 1.807 times as much butter-fat as Nora, the difference in yield between the two cows was 252.27 lb. of butter-fat or 294.31 lb. of butter per year. This at 16 cents per pound, which is the average value of butter before being made up, would amount to $47.09 per year. Supposing that the cows would yield in this ratio for six years, from the age of four to ten, which is a conservative estimate, Rose would produce $282.54 worth of butter more than Nora on exactly the same kind and quantity of feed:

RECORD OF THE TWO COWS FOR ONE YEAR COMPUTED ON A LIKE FEED BASIS

	ROSE	NORA	DIFFERENCE
Reduced to a like feed basis the amount Nora would have produced had she eaten the same as Rose:			
Total digestible dry matter consumed in pounds.	6477.92	6477.92	—
Total yield of milk, in pounds.	11,329.00	8121.60	3207.40
Total yield of butter-fat, in pounds	564.80	312.53	252.27
Total yield of butter, in pounds	658.90	364.62	294.28
Total value of butter at 16¢ per pound.	$105.43	$58.34	$47.09

"In this comparison Rose was at a disadvantage in two ways. She was nine years of age and on the down grade of life while Nora was just in her prime. Rose was bred November 5, 1899, while Nora was not bred until after the experiment closed. Had it not been for these two hindrances Rose would doubtless have made even a better record than she did.

"While there is a vast difference in the profit derived from the two cows in this experiment, the difference is by no means phenomenal, as greater differences than here cited may frequently be found among cows in the same herd, for the cow Nora, the poorer of the two, was in reality an exceptionally good cow, producing 348 lb. of butter in a year which is nearly three times the average yield (130 lb.) of cows in the United States and almost one-half more than the average yield (250 lb.) of profitable cows in Illinois. Had Rose been compared with a really poor cow, such as may be found in nearly all dairy herds, there would have been a much greater difference in profit in favor of Rose; for she gave nearly five times as much as a profitable cow for Illinois."

In the cases of variation mentioned, the individuals compared have belonged to different breeds or have been unrelated. It is also true that animals which are from parents of the same blood lines may show wide variations. In the dairy herd belonging to the University of Missouri, there were at one time nineteen daughters of the Jersey bull Minette's Pedro. Many of the mothers of these cows were from another bull and all were of similar breeding. The conditions surrounding these cows were alike; their dams, grand dams, and great grand dams were all similarly bred and every cow of the nineteen was sired by Minette's Pedro. These cows should have exhibited some uniformity in the development of dairy qualities. The table records the annual production of milk and butter, and shows rather wide variations in the productive capacity of the different cows:

RECORDS OF THE DAUGHTERS OF MINETTE'S PEDRO[1]

	Number Lactation Periods	Average Lbs. Milk	Average Lbs. Fat
Pedro's Ramaposa	3	6750	365.8
Pedro's Elf	3	2225	109.4
Pedro's Alphea Elf	5	6151	309.8
University May	3	4723	227.0
Columbia Huguita	5	6322	273.1
Pedro's Daisy Bate	2	3456	193.7
Missouri Daizie	1	4910	205.5
University Daizie	4	7746	405.6
University Stella	3	5336	273.7
University Elf	4	5053	247.3
University Belle	3	4960	223.8
Pedro's Grace Briggs	3	4909	287.9
Pedro's Matron	3	6582	355.9
Miss Missouri	3	6844	331.0
Pedro's Emily Harris	3	5271	238.5
Pedro's Estella	2	8807	462.1
Pedro's Alphea Ward	1	4728	267.0
Pedro's May Hubbard	4	4073	184.9
Pedro's Virginia Meredith	3	5776	320.6

195. New characters originate in the germ-plasm. — In the preceding discussion of the possibility of the inheritance of acquired characters, we have followed closely Weismann's definition of acquired characters. It must be admitted, however, that in the discussion of this subject by many students of heredity, the use of the term has not been confined strictly to Weismann's interpretation. Many biologists would include under the discussion of this subject all acquired characters, regardless of whether they may have originated through

[1] Eckles, Missouri Experiment Station, Research Bulletin, No. 2, p. 108.

environmental influences acting upon the soma-cells or from variations directly affecting the germ-plasm itself.

Many new characters appearing in plants and animals cannot be traced to environmental influences acting upon the soma-cells. Many characters seem to arise independently of external causes. They undoubtedly have their origin in the germ-plasm itself. Such variations are fundamentally different from the characters which are acquired by the soma-cells as the result of environment, use or disuse, disease and mutilations.

The germ-cell which contains within its own substance the materials needed for giving direction to the development of the new individual is the result of the union of two other germ-cells from individuals which may represent widely different characters. It is impossible that the new individual arising from such a germ-cell shall possess characters identical with either parent. The offspring represents a certain amount of variation which had its origin in the germ-cell itself.

The mechanism of reproduction, including the maturation and reduction of the germ-cells and the union of the chromosomes, provides ample opportunity for new combinations of characters which may profoundly change the whole physiological history of the offspring.

196. Mutilations. — Many examples of mutilations and their supposed transmission from parent to offspring have given to advocates of the belief in the transmission of acquired characters many of their most interesting examples. Various investigators have cut off the tails of mice for many generations with a view to investigating the result of such mutilation upon its transmissibility. Cope, Mantegazza and Rosenthal cut off the tails of mice for eleven generations, Bos for fifteen, and Weismann for nineteen generations, but in no single case was there the slightest evidence that this form of mutilation was transmitted.

The tails of sheep have been cut off for many hundred years by shepherds, but tails reappear regularly with the normal number of vertebræ and without diminution in length.

A large number of examples have been given of cats whose tails have been removed and who have later transmitted to their offspring a tendency to short tails. In the Eiffel, the peasants shorten the tails of cats. It is reported by Tietz that cats with defective tails are common in this region. Many similar examples are reported. In this and most other examples of the inheritance of acquired characters, there is no evidence that the artificial shortening of the tails is the direct cause of the atrophied tails observed in the kittens. Such defective tails are not uncommon in races of cats with normal tails. It must also be remembered that there are tailless breeds of cats such as the Manx and Japanese breeds, and the admixture of such breeds might be sufficient to explain the observed variations.

The cattlemen of the Nile Valley have for an unknown period of time caused the horns of cattle to grow in curious spiral forms, but there is no evidence that such deformities are transmitted to the offspring.[1]

197. The Brown-Sequard experiments.[2] — The most frequently quoted and credible scientific experiment conducted for the purpose of causing somatic modifications and observing their transmission from parent to

[1] Hartman, " Die Haussäugethiere der Wildländer," *Ann. Landwirthsch.*, Berlin, 1864, p. 28.
[2] Romanes, " Darwin and after Darwin," vol. II, chap. IV.

offspring are the famous Brown-Sequard experiments with guinea pigs. From 1869 to 1891, Brown-Sequard cut the sciatic nerve of the leg or the spinal cord in the dorsal region, causing an abnormal nervous condition resembling the symptoms of epilepsy. These animals when allowed to breed produced offspring, many of which were epileptic like the parents. Similar results were later secured by Westphal, Dupuy, Obersteiner and Romanes. This interesting investigation has been promptly accepted by the special advocates of the transmission of acquired characters as fulfilling the oft repeated demand for direct evidence of the inheritance of somatic modifications.

In discussing the results it must not be forgotten that the mutilation was never transmitted, but only the epileptic state resulting from the mutilation. The results from this type of mutilation were very diverse. According to Romanes, the epileptic condition was rarely transmitted. Brown-Sequard admitted that certain particular results were exhibited in only one or two per cent of cases. If this mutilation had actually influenced the germ-plasm in such a way as to add to its fundamental constitution the determiners essential for the development of the new characters, then surely we might expect a larger part of the offspring to be affected with the acquired character.

Max Sommer in 1900 repeated the Brown-Sequard experiments, but failed to confirm the conclusion that this experiment had proven the existence of acquired characters. "As regards the hereditary transmission of epilepsy in guinea pigs," says Sommer, "or of other accidentally acquired pathological symptoms — *e.g.* defects in the toes — we have not been able to confirm the experiments of Brown-Sequard and Obersteiner; and we do not think that these can any longer serve as a support to the doctrine of the inheritance of acquired characters."[1]

198. Causes of variation. — The nature of variation is still obscure. The fundamental causes are not easily determined. "Our ignorance of the laws of variation is profound," says Darwin.

The results of the investigations in cytology have given a more reasonable basis for understanding the subject of variation, but it has not yet given us a wholly satisfactory knowledge of the causes of variation. Bateson holds that we are yet far from a satisfactory explanation of the real nature of variation. He has concluded that "Inquiry into the causes of variation is, in my judgment, premature." We are, however, able to recognize and classify certain apparent causes of variation. Such classification recognizes causes of variation as external and internal.

Davenport[2] has further classified the internal causes of variation as: "1. Internal influences affecting primarily the individual, and 2. internal influences affecting the race as a whole."

199. Cell division a cause of variation. — Every animal is the product of the union of the germ substance of two other animals. The union of the germ-cells is a union of the characters of the parents. This combination of the germinal matter of the two parents results in a rearrangement of some of the characters, and these may vary materially from the original characters of the parents. Weismann calls this mixing of the germ-plasm amphimixis. He is of the opinion that sexual reproduction by cell union and cell division is nature's plan for increasing variation.

[1] Thomson, "Heredity," pp. 230–236.
[2] Davenport, "Principles of Breeding," p. 155.

New characters may arise or old ones be lost through accidents to the germ substance during the processes of cell division incident to reproduction and growth. It is conceivable that a chromosome bearing within its material substance a character or set of characters may be lost or destroyed at some stage of the complicated processes which eventually result in the formation of a new germ-cell. If such a thing occurs, it must influence greatly the ultimate characters of the individual. It is also apparent that a fundamental change in the germinal material may influence not alone the resulting individual but the race or breed. Sudden marked variations may and do often occur, and these may become the beginnings of new races. These sudden variations are called mutations by De Vries, and the mutation theory of evolution is regarded as one of the most important advances since Darwin. A fuller discussion of mutations will be found on another page.

It is difficult to differentiate between those variations which are merely the result of the action of environment on the soma- or body-cells and those variations which are the result of a fundamental change in the constitution of the germ substance. The former are generally temporary and affect only the individual, but do not influence the germ sufficiently to cause the same variation to appear in the offspring. The highly improved types of domestic swine if permitted to run wild in the woods lose their rounded full-fleshed form and assume much the appearance of the unimproved " razor-back." Their appearance is so changed by this treatment that the most skillful judge of swine would be greatly deceived. The " razor-back," on the other hand, has subsisted for generations upon the mast of the forest. This has involved much exercise and the ability to live on a scant supply of food at certain seasons. These conditions have resulted in changing the form of the body. The neck and jaws are larger and well muscled. The back is sharp, legs longer and ribs flatter than in the improved forms. If the young pigs of these unimproved swine are placed under conditions where they are supplied with an abundance of nutritious food, they approach somewhat the well-rounded form of the improved type. In each of these cases the environment has resulted in causing a distinct variation from the parent form. This variation can be easily observed, it can be measured. But whether this environment has influenced in any way or how much it has influenced the elemental carriers of heredity in the germ, it is impossible to do more than conjecture. It cannot be accurately measured. Biologists are generally agreed that in a case of this kind the germ is not fundamentally changed.

200. Influence of use and disuse in causing modifications. — The constant use of an organ in the performance of work will modify the organ in accordance with the work performed. Any organ of the body that is not used may atrophy. Changes resulting from excessive use of the various parts of the body may be so extensive as to appear almost as new characters. If modifications resulting from food, climate or mutilations are transmitted, the changes resulting from the constant use of an organ should be transmitted with equal or greater force.

Lamarck's theory of evolution was largely based upon the supposed adaptation of the various organs of the body to their environment, and that such adaptations were readily transmitted. Thus the giraffe, forced by conditions to feed upon the leaves of trees, gradually extended his neck, which became longer, and this increase in length was transmitted from generation to generation. Wading birds, feeding in the shallow water along the shore, gradually waded deeper and deeper into the water. Their legs became longer, and the additional length gained by each generation was transmitted. The long tongue of the ant-eater, of woodpeckers, and humming birds, was developed in a similar manner. The rudimentary eyes of subterranean animals and fish in caves is another supposed example of the loss of an organ through disuse.

Among domestic animals, there are numerous examples of a high degree of development of organs through continued exercise. The milking function in the dairy cow can undoubtedly be greatly improved in any individual by skillful exercise and use. The training of running and trotting horses has resulted in very greatly increasing the ability of an animal in those particular types of speed. Are these modifications, the result of use or disuse, transmitted by heredity? Such inheritance would be of the very greatest importance to the breeder of domestic animals. In answer to this question no direct proof has been offered that characters acquired by exercise or lost by disuse are actually transmitted by heredity.

201. Importance of causes of variation to the breeder of domestic animals. — While the researches of biologists have led them to believe that the germ-plasm is very stable and its character not easily changed by the environment of the body, it is nevertheless true that breeders of the domestic animals have long believed that the amount and kind of food, climate and training which animals receive has an influence not only upon the individuals benefiting by or suffering from such environment, but likewise may have a profound influence upon their posterity. The breeder of beef cattle believes that the offspring of parents which are kept in good or even in fat condition are more apt to possess a tendency to fatten readily than the offspring of parents kept in very thin condition. The breeder of trotting horses prefers to use in his stud a stallion that has a record and mares that have benefited from severe training.

In this case the biologist is probably in the main correct in his conclusions from the standpoint of inheritance. But it is also true that the breeder of beef cattle is right in maintaining his beef animals on a high plane of nutrition, not because this will materially affect the germ-plasm, but because such treatment gives the breeder an accurate measure of the beef-producing characters of his breeding animals. How can the breeder know that a particular bull or cow possesses the ability to lay on fat rapidly unless he actually tests the animal? The breeder of trotting horses likewise cannot judge accurately from an external examination of a horse how fast he can trot. He must be trained and his full speed developed.

Such treatment on the part of the breeder is not for the purpose of changing the hereditary capacities of an animal, but for the purpose of aiding selection. Animals so treated that do not come up to the standard set by the breeder are eliminated. The desirable animals are preserved and encouraged to reproduce.

202. Germinal variations. — The term variation has suffered from careless use, and such use has led to some confusion of ideas. Differences appearing in the offspring may be due to variations in the germ, or they may be due to the influence of environment. What the animal actually is depends upon the constitution of the germ. The offspring may be exactly like the parent in the constitution of the germ substance from which each has been developed, but they may appear to be different. Such differences may be due to a change in the environment which, acting upon the organism, may have modified the apparent character of the individual. Such changes are not variations in the true sense, but rather modifications. It is not always possible to distinguish readily between changes which are merely modifications and variations which are due to fundamental changes in the germ. To the practical breeder, it is in the highest degree important that this distinction between mere modifications due to environment and germinal variations due to a change in the constitution of the germinal substance be clearly recognized. The latter variations are strongly transmitted by heredity; the former are not transmissible. Domestic animals kept under the same conditions often exhibit wide variations, and these are often germinal and consequently inherited. Those variations or, more properly, modifications which appear in individuals and are the result of environment are of little significance to the breeder. If the breeder of speed horses confined his selection solely to those horses that had been trained, he might not secure the sum total of those characters in the fundamental constitution of the animal which represent the highest capacity for speed. It is true that among horses of similar ancestry the training and development is the most accurate index of the capacities which they have inherited. But a horse that has not been trained and hence modified by environment may actually possess through inheritance a greater capacity for speed. The latter horse may show less speed than the horse that has been carefully developed, but he will be a better breeder. Let us assume for example that we have two horses under consideration, of the same ancestry. One horse has through germinal variation been endowed with an ability to trot or run at a certain speed without training. The other horse cannot attain the same speed except as the result of long and careful training. They have attained the same rate of speed, but one has acquired this speed through the influence of environment and this increased speed becomes, therefore, a mere modification. The other horse owes his ability to go fast to a variation in the fundamental constitution of the germ substance. The latter horse will be the better breeder because germinal variations are transmitted, and modifications which result from the influence of environment are apparently not transmitted.

CHAPTER XI

IN-BREEDING

THE breeder of domestic animals is frequently confronted with the problem of in-breeding. If in-breeding is not followed by injury, it would often be a convenient method of improvement. The value of a proven and tested sire is so great that if he could be safely mated with his own offspring it would be of great economic advantage to the breeder. If, on the other hand, positive advantages follow the mating of closely related animals, the breeder should know what these advantages are and how and when they may be most certainly realized.

203. Definitions. — In-breeding has been variously designated as close-breeding, consanguineous breeding, in-and-in-breeding, inter-breeding and incestuous breeding. The term in-breeding is used to indicate the mating of animals which are near of kin or closely related. The degree of relationship which it is proper to designate as in-breeding is a matter of some disagreement. Stonehenge, for example, has defined in-breeding as " The pairing of relations within the degree of second cousins, twice or more in succession." Randall would restrict the application of the term to " animals of precisely the same blood as own brother and sister."

It would be very desirable if the term " in-breeding " could be limited in its application as suggested by Morgan.[1] "For species with separate sexes the term 'in-breeding' is used to express either the union between brothers and sisters or between offspring and parent, in one or more generations." Unfortunately the literature on the subject of in-breeding has not placed such narrow limitations on the term.

It must be recognized that there are different degrees of in-breeding. Animals may be closely in-bred as, for example, results from the mating of parent and offspring, or brother and sister. The union of more distant relationships, as third or fourth cousins, would not be expected to show the same good or bad results in the offspring as the more closely related parents. In the literature of the subject, the discussions generally have reference to the most intensive forms of close-breeding. The results, good or bad, therefore, are those which may be expected to follow the most intensive in-breeding. After all, the real question of importance for the practical breeder to answer is not whether any form of in-breeding should be practiced, but to what extent it may be practiced and its known advantages become realized. In a sense, practically every registered improved breed to-day is the result of a certain amount of in-breeding. David Starr Jordan has calculated that if no in-breeding of any degree had taken place in the human race, each person born in the thirtieth generation from William the Conqueror would have had 8,598,094,592 living ancestors at the time when the Conqueror was alive.

204. Advantages claimed for in-breeding. — The mating of animals having the same parentage, has resulted in certain definite advantages to the race or breed. These results are as easily demonstrable as are the results from

[1] Morgan, "Experimental Zoölogy," 1907, p. 186.

any other system or method of breeding. The particular beneficial result most commonly claimed is that in-breeding is the quickest method of fixing and perpetuating a desirable character. Closely related animals are most likely to possess the character sought, and mating animals having the qualities which the breeder particularly desires to perpetuate in the breed is the most natural method of accomplishing his purpose. In-breeding tends to intensify the good qualities which the breeder is striving to make dominant. It does not cause new and desirable characters to appear, but is merely a method of making the greatest possible use of such characters.

Thus in-bred animals are strongly prepotent. They possess to an unusual degree the power of fixing their qualities upon their offspring. This is manifestly the most important characteristic in a highly improved breeding animal. Next to the possession of the highly improved characters which make the domestic animals useful to man, their ability to transmit those qualities is most important. In-breeding is one certain means of developing the prepotency of animals.

It is a fundamental principle of breeding that the smaller the number of qualities selected for improvement by the breeder, the more rapid and certain will be his progress in the improvement of the breed. In-breeding tends to reduce the number of characters, simplify the breeding operations, and thus makes more certain the continued reappearance of the valuable characters in succeeding generations.

205. Bad results from in-breeding. — In recounting the well-known benefits which follow intelligent in-breeding, it is not intended to convey the impression that in-breeding results only in success. The biological processes which result in simplifying the germ-plasm and intensifying the powers of transmission act impartially on all characters alike, bad as well as good. Lurking tendencies to evil may become strengthened along with the good, and thus be more strongly transmitted than before.

Such a result cannot always be foreseen and hence when breeding closely related animals, there is always the risk that we will produce offspring which are not only more prepotent in respect to the good qualities we are seeking to develop and perpetuate, but we may at the same time bring about the same result in connection with the bad qualities.

But aside from the bad results following in-breeding which may be ascribed to the simplifying of the germ-plasm and the intensifying of the tendencies to evil, it has long been held by many eminent biologists and by practical breeders that certain definite evil results always follow long-continued in-breeding. The most important of these necessary evils are loss of vigor, decreased fertility and diminished size.

206. Decreased fertility and vigor from in-breeding. — Many breeders believe that continuous in-breeding results in a loss of fertility. It is admitted that most other qualities may be advantageously improved by close-breeding, but that the quality of fertility is an exception. This belief is firmly implanted in the minds of the greater number of breeders and of many biologists. The basis for such a belief is found in the results of certain specific investigations and the general experience of breeders. Fertility is a character of prime importance in the domestic animals. This character is undoubtedly subject to the same general laws of transmission as are all other hereditary qualities. We have seen how in-breeding may be used to intensify and fix the other desirable characters of a breed, and incidentally greatly increase their prepotency. If an animal is possessed of the quality of fertility to an unusual degree, why may not in-breeding be employed to increase fertility as well as to improve the qualities of speed in horses or of early maturity in meat-producing animals? Let us answer the question by an examination of the available data. Is in-breeding *per se* specifically injurious to the fertility of plants and animals? If in-breeding is injurious at all, how serious is the injury and how far can the breeder take advantage of the known good results without sacrificing the important quality of fertility? The data available for answering these questions are to be found in the practical experience of breeders and the results from carefully planned experiments where all other factors have been eliminated excepting only the factor of in-breeding.

207. Darwin's researches. — The greatest single contribution to the subject of in-breeding was made by Darwin. Recognizing the advantages of close in-breeding in fixing desirable characters and admitting that these advantages may outweigh possible injury, he brings forward an array of examples of the injurious effects of in-breeding which are convincing. His conclusions are best stated in his own words:

[1] "That any evil directly follows from the closest inter-breeding has been denied by many persons; but rarely by any practical breeder; and never, as far as I know, by one who has largely bred animals which propagate their kind quickly. Many physiologists attribute

[1] Darwin, "Animals and Plants under Domestication," vol. II, p. 94.

the evil exclusively to the combination and consequent increase of morbid tendencies common to both parents; and that this is an active source of mischief there can be no doubt. It is unfortunately too notorious that men and various domestic animals endowed with a wretched constitution, and with a strong hereditary disposition to disease, if not actually ill, are fully capable of procreating their kind. Close inter-breeding, on the other hand, often induces sterility; and this indicates something quite distinct from the augmentation of morbid tendencies common to both parents. The evidence immediately to be given convinces me that it is a great law of nature, that all organic beings profit from an occasional cross with individuals not closely related to them in blood; and that, on the other hand, long-continued close inter-breeding is injurious."

Darwin's conclusions are based upon a very large number of observations. Experienced breeders who are accurate observers, such as Sir J. Sebright,[1] Andrew Knight and Herman von Nathusius, all agree as to the certain injury which always follows long-continued in-breeding.

Darwin's investigations led him to believe that while many species of plants and animals are hermaphroditic and hence self-fertilizing, and these might be presumed perpetually to fertilize themselves, yet he failed to find a single species in which nature had provided structures which insured self-fertilization. On the other hand, he found innumerable instances in which nature had provided special structures for the sole apparent purpose of insuring cross-fertilization and thus preventing perpetual in-breeding.

[1] Sebright, "The Art of Improving the Breed," 1809.

208. In-breeding cattle. — Bakewell's [1] (1725–1795) phenomenal success in the rapid improvement of horses, cattle and sheep was possible only because he utilized to the fullest extent the method of mating animals of the closest possible relationship, not because they were closely related but because they possessed the particular qualities desired. By in-breeding he was able to "simplify" the germ-plasm and bring about a homozygous condition of these particular characters. The available breeding records of the activities of Robert and Charles Colling, Thomas Bates and the Booths are eloquent in their testimony of the fact that great progress was achieved from intelligent in-breeding.

The Shorthorn bull, Duke of Airdrie (12,730),[2] traces through five or six generations to but six animals famous in the early history of the Shorthorn breed. The six animals all trace back through five or six generations to one bull Favourite, himself the son of half-brother and sister. Says Darwin,[3] "But the Shorthorns offer the most striking case of close inter-breeding; for instance, the famous bull Favourite (who was himself the offspring of a half-brother and sister from Foljambe) was matched with his own daughter, granddaughter, and great-granddaughter; so that the produce of this last union, or the great-great-granddaughter, had 15-16ths, or 93.75 per cent of the blood of Favourite in her veins. This cow was matched with the bull Wellington, having 62.5 per cent of Favourite blood in his veins, and produced Clarissa; Clarissa was matched with the bull Lancaster, hav-

---

[1] Youatt, "Cattle," p. 199.

[2] A valuable discussion of in-breeding among early breeders is to be found in Miles' "Stock Breeding," pp. 137–189. Also Huth, "The Marriage of Near Kin," pp. 242–292.

[3] Darwin, "Animals and Plants under Domestication," p. 96.

ing 68.75 of the same blood, and she yielded valuable offspring. Nevertheless Collings, who reared these animals, and was a strong advocate for close-breeding, once crossed his stock with a Galloway, and the cows from this cross realized the highest prices. Bates's herd was esteemed the most celebrated in the world. For thirteen years he bred most closely in-and-in; but during the next seventeen years, though he had the most exalted notion of the value of his own stock, he thrice infused fresh blood into his herd: it is said that he did this, not to improve the form of his animals, but on account of their lessened fertility."

The opinion of a great breeder who has practiced very close in-breeding for eighty years and who has also been noted for the general success of his breeding operations and the high quality of his cattle, so much so as to have been called one of the founders of the breed, is of great interest in this connection. Such a breeder was Price of England. He says,[1] "My herd of cattle has, therefore, been bred in-and-in, as it is termed, for upward of eighty years, and by far the greater part of it in a direct line, on both sides, from one cow now in calf for the twentieth time. I have bred three calves from her, by two of her sons, one of which is now the largest cow I have, possessing also the best form and constitution; the other two were bulls, and proved of great value, thus showing indisputably that it is not requisite to mix the blood of the different kinds of the same race of animals, in order to keep them from degenerating."

209. The Chillingham cattle. — The wild white cattle of Great Britain are believed to be the only living pure descendants of the original wild cattle of the British Is-

---

[1] *Farmers' Magazine*, 1841, vol. XIV, p. 50.

lands. These cattle have been kept pure in various private parks of Great Britain and have for many hundred years been subjected to conditions which compelled extensive in-breeding. These cattle have not perished, they are not weak in constitution and are not decreasing in size. This example is often quoted as an argument in favor of in-breeding. Darwin's critical analysis of this classic example leaves us still in doubt as to its value as a demonstration of the beneficial results which may be expected to follow long-continued in-breeding.

[1] "The half-wild cattle," says Darwin, "which have been kept in British parks probably for 400 or 500 years, or even for a longer period, have been advanced by Culley and others as a case of long-continued inter-breeding within the limits of the same herd without any consequent injury. With respect to the cattle at Chillingham, the late Lord Tankerville owned that they were bad breeders. The agent, Mr. Hardy, estimates (in a letter to me dated May, 1861) that in the herd of about 50 the average number annually slaughtered, killed by fighting and dying, is about 10, or one in five. As the herd is kept up to nearly the same average number, the annual rate of increase must be likewise about one in five. The bulls, I may add, engage in furious battles, of which battles the present Lord Tankerville has given me a graphic description, so that there will always be rigorous selection of the most vigorous males. I procured in 1855 from Mr. D. Gardner, agent to the Duke of Hamilton, the following account of the wild cattle kept in the Duke's park in Lanarkshire, which is about 200 acres in extent. The number of cattle varies from 65 to 80; and the number annually killed (I presume by all causes) is from 8

[1] Darwin, "Animals and Plants under Domestication," p. 97.

to 10; so that the annual rate of increase can hardly be more than one in six. Now in South America, where the herds are half-wild, and therefore offer a nearly fair standard of comparison, according to Azara the natural increase of cattle on an estancia is from one-third to one-fourth of the total number, or one in between three and four, and this no doubt applies exclusively to adult animals fit for consumption. Hence the half-wild British cattle which have long been inter-bred within the limits of the same herd are relatively far less fertile. Although in an unenclosed country like Paraguay there must be some crossing between the different herds, yet even there the inhabitants believe that the occasional introduction of animals from distant localities is necessary to prevent ' degeneration in size and diminution in fertility.' The decrease in size from ancient times in the Chillingham and Hamilton cattle must have been prodigious, for Professor Rutimeyer has shown that they are almost certainly descended from the gigantic *bos primigenius*. No doubt this decrease in size may be largely attributed to less favorable conditions of life; yet animals roaming over large parks, and fed during severe winters, can hardly be considered as placed under very unfavorable conditions."

210. Deer in parks. — In many English parks fallow deer have been kept for many decades, and in-breeding must often result. An investigation by Darwin disclosed the fact that the managers of such parks found it necessary to introduce new blood to improve the size, constitution, vigor and prevent the taint of " rick back," which follows too close breeding.

211. In-breeding among pigs. — The evil results from in-breeding are naturally more quickly apparent among animals which produce large numbers of young at a birth and have a comparatively short period of gestation. Domestic swine fulfill these conditions admirably and are therefore most valuable material for breeding experiments.

[1] "Mr. J. Wright, well known as a breeder, crossed the same boar with the daughter, granddaughter, and great-granddaughter, and so on for seven generations. The result was, that in many instances the offspring failed to breed; in others they produced few that lived; and of the latter many were idiotic, without sense, even to suck, and when attempting to move could not walk straight. Now it deserves especial notice, that the two last sows produced by this long course of inter-breeding were sent to other boars, and they bore several litters of healthy pigs. The best sow in external appearance produced during the whole seven generations was one in the last stage of descent; but the litter consisted of this one sow. She would not breed to her sire, yet bred at the first trial to a stranger in blood. So that, in Mr. Wright's case, long-continued and extremely close inter-breeding did not affect the external form or merit of the young; but with many of them the general constitution and mental powers, and especially the reproductive functions, were seriously affected."

Nathusius reports that as a result of closely in-breeding Yorkshire swine for three generations the offspring were weak in constitution and their fertility was impaired.

212. In-breeding sheep. — The American Merino sheep is a remarkable example of what intelligent breeding and

---

[1] Darwin, "Animals and Plants under Domestication," p. 101.

Also *Journal of Royal Agricultural Society*, 1846, vol. 7, p. 205.

selection can accomplish in the improvement of the domestic animals. The first importation of Spanish Merino sheep to America was made in 1815. The average weight of fleece of these sheep at that time was three or four pounds a head. This average was increased by American breeders until, in 1880, the average fleece from selected flocks was fifteen pounds a head, and single individuals were produced which sheared as high as thirty-five pounds. Plumb[1] reports that the heaviest fleece on record weighed 44 pounds and 3 ounces and was taken from a two-year-old ram at the public shearing of the Vermont Sheep Shearing Association. But the finest specimens of the American Merino breed were the result of in-breeding. " Mr. Atwood bred his entire flock from one ewe, — and thus, after being drawn beyond all doubt from an unmixed Spanish Cabana, they have been bred in-and-in, in the United States, for upward of sixty years."[2] " The ram Gold Drop for which Mr. Hammond refused twenty-five thousand dollars,"[3] was closely in-bred.

213. In-breeding dogs. — Many examples of in-breeding among dogs are mentioned in the literature of breeding, but unfortunately the records of these cases are very incomplete and many are of doubtful scientific value. The author has had an opportunity to examine somewhat carefully the results of long-continued in-breeding among fox terriers. Arthur Rhys, the herdsman at the University of Missouri, has practiced very close in-breeding of fox terriers for nine generations. Daughter No. 1, from wholly unrelated parents, was bred back to Designer, her

---

[1] Plumb, "Types and Breeds of Farm Animals," p. 349.
[2] Randall, "Practical Shepherd."
[3] Miles, "Stock Breeding," p. 150.

own sire, producing two litters of six and seven each. Female No. 2, a daughter of No. 1, was bred to Designer, her own sire, who was also her grandsire. She produced a litter of eight. Female No. 3, a daughter of No. 2, was bred again to Designer, her own sire, who was also her grandsire and great-grandsire. Female No. 4 was again bred to Designer, her own sire, who was also her grandsire, great-grandsire and great-great-grandsire. She produced a litter of eight. Thus in turn Females Nos. 5, 6, 7 and 8 were bred to their own sire, Designer. Female No. 8, resulting from this long-continued in-breeding, was bred to her own son, and from the litter resulting a brother and sister were selected and in-bred. Mr. Rhys states that, "I see no evidence of decrease in size of bone, in constitution or in fertility as a result of my experience in in-breeding fox terriers." The only peculiarity which has been observed in the later generations as compared with the original animals is a slight lack of courage or "nerve" in the later animals. A normal fox terrier never flinches in the face of sudden danger.

The illustration, Plate XIV, lower, represents Designer II, at four months old, one of the seven dogs in the eighth generation of continuous in-breeding. Every litter for eight generations was sired by Designer I. The illustration, Plate XIV, upper, pictures the dog Dispatcher of the ninth generation, son of own brother and sister.

214. Cornevin's[1] **experiments.** — The French breeder and author, Cornevin, practiced in-breeding with swine, cattle and sheep for considerable periods without injury. He in-bred Jersey cattle for seven years, Hollander cattle for twelve years, and Merino sheep eleven years without observing any evidences of degeneracy. His experi-

---

[1] Cornevin, "Traité de Zootechnie Generale" (1891).

ments with swine were unfavorable to the practice of in-breeding. According to Cornevin, among pigeons it is the rule for brother and sister to mate. The same is also generally true of ducks, geese, guinea fowls and swans. After eleven years of in-breeding pigeons and geese, he was unable to observe any changes in color, weight or fecundity which could be ascribed to in-breeding.

Georg Wilsdorf,[1] the German authority, has found by investigation that most pure breeds have resulted from in-breeding. He says, "In our studies of the history of various breeds, we next made the astonishing discovery that the best living individuals belonged to families which, when their pedigrees were traced, were found all to come from a single family — often from a single individual. By way of illustration I might cite the Hanoverian halfbloods, which we know particularly through the studies of de Chapeaurouge and Grabensee to have come almost altogether from three stallions, of which Norfolk has hitherto had the greatest influence on the breed — an influence that is increasing all the time. Researches into the swine breeding of the Visselhövede district, and into that of Hildesheim in Bavaria, have shown that in each case a single boar was the ancestor of various valuable families, to-day widely scattered. And Hoesch of Neukirchen has found that his valuable strain of swine is principally due to the blood of a single early boar Richard."

"The modern science of breeding, however, stands firm in its belief that for the production of definite types for special purposes in-breeding is the quickest and most certain method of procedure, and all great breeders who

[1] *Journal of Heredity*, March, 1915, pp. 110–111.

work toward any particular goal depend largely on in-breeding, knowingly or unknowingly." [1]

215. Weismann's and Von Guaita's experiments. — Weismann [2] in-bred mice for twenty-nine generations. As shown in the following table, there was a constant and fairly uniform decrease in fertility from the first to the last generation:

1 to 10 generations; 1345 young; 219 litters; avg. per litter 6.1
11 to 20 generations; 252 young; 62 litters; avg. per litter 5.6
21 to 29 generations; 124 young; 29 litters; avg. per litter 4.2

The average number of young to a litter decreased from 6.1 in the first ten generations to 4.2 in the last ten generations. Whether this decrease is due directly to the specific action of in-breeding on the quality of fertility, or whether it simply represents an intensification of an innate tendency to low fertility which existed in this particular strain, it is not possible to determine. Von Guaita, working with the same strain of mice and beginning with the last generation (29th) bred by Weismann, obtained the following results:

1st and 2d generations, avg. per litter, 3.5
3d and 4th generations, avg. per litter, 3.6
5th and 6th generations, avg. per litter, 2.9

There is here a clear loss of fertility from an average of 6.1 to a litter to 2.9 to a litter in 35 generations, traceable to in-breeding.

[1] Georg Wilsdorf, "Tierzuchtung," 1912.
[2] "Berichte der Naturforschenden Gesellschaft zu Freiburg," 1900.
See Morgan, "Experimental Zoölogy," p. 188.

216. Researches of Ritzema Bos.

An interesting experiment with in-breeding white rats for thirty generations is reported by Ritzema Bos.[1] Beginning with an albino female mated with a wild rat, the first mating resulted in twelve offspring. Seven of this litter were bred to a white male but unrelated. The resulting offspring were closely in-bred for six years. The matings were brothers to sisters and parents to offspring. The average size of the litters for twenty generations and covering a period of four years remained practically constant. The last ten generations born during the last two years of the experiment showed a decided decrease in the fertility of the matings. The average size of litters by years is shown in the table.

DECREASED FERTILITY DUE TO IN-BREEDING. (Bos)

1887	1888	1889	1890	1891	1892
$7\frac{1}{4}$	$7\frac{1}{2}$	$7\frac{14}{15}$	$6\frac{11}{18}$	$4\frac{7}{12}$	$3\frac{1}{5}$

Not only the average size of the litters decreased in thirty generations from $7\frac{1}{2}$ to $3\frac{1}{5}$, but the number of matings which were sterile increased greatly from the first to the last generations, as shown in the following table:

MATINGS WHICH PROVED STERILE. (Bos)

1887	1888	1889	1890	1891	1892
0	2.63	5.55	17.39	50	41.18

There is evidence also in this experiment that the constitutional vigor of the offspring of parents which were the result of long-continued in-breeding was materially injured. The time of the appearance of weakness

---

[1] Ritzema Bos, "Biol. Centralb.," XIV, 1894. *See also* Morgan, "Experimental Zoölogy," p. 188.

in the offspring was coextensive with the decline in fertility.

The comparative mortality of the young increased rapidly, as shown in the following table:

INCREASED RATE OF MORTALITY DUE TO IN-BREEDING. (Bos)

1887	1888	1889	1890	1891	1892
Per cent	Per cent	Per cent	Per cent	Per cent	Per cent
3.9	4.4	5.0	36.7	36.4	45.5

The bad effects from in-breeding were more noticeable when brother and sister were mated than when parents were mated with offspring. Of the matings between parent and offspring 21.4 per cent were sterile, while those between brother and sister were 36 per cent infertile. A tendency to decrease in size is indicated by the record of weights of different generations. Full-grown male rats of the first generation weighed 300 grams each. In the tenth generation the weight had decreased to 275 grams and at the end of six years the weight of rats had declined to 240 grams.

217. The Wistar Institute experiments. — Extensive in-breeding experiments by Helen Dean King[1] at the Wistar Institute, Philadelphia, with white rats seem to have resulted in disproving the theory that in-breeding is always and necessarily followed by evil results. In this investigation white rats have been as closely in-bred as possible for twenty-two generations. More than 10,000 in-bred rats have been observed during a period of seven years. The original stock was two pairs of albino rats. Each pair was used as the foundation for a series of in-breeding experiments. In each generation

[1] *Journal of Heredity*, vol. 7 (1916), p. 70.

the best rats from the standpoint of size, vigor, fecundity and general quality were carefully selected.

The actual results of this investigation after twenty-two generations of in-breeding are that the male in-bred rats are about fifteen per cent heavier and the female in-bred rats about three per cent heavier than stock rats. The size of litters among the stock rats has been seven, while among the in-bred rats the litters have increased and now average seven and four-tenths each. The thrift and vigor of in-bred rats in these experiments was apparently not injured by in-breeding. "The results so far obtained with these rats," says Dr. King, "indicate that close in-breeding does not necessarily lead to a loss of size or of constitutional vigor or of fertility, if the animals so mated come from sound stock in the beginning and sufficient care is taken to breed only from the best individuals."

218. In-breeding Berkshires by Mr. Gentry. — The remarkable success which may follow the practice of in-breeding when intelligently conducted by a skillful breeder is shown in the experience of N. H. Gentry of Sedalia, Missouri. Probably no American breeder has been so successful in developing all those desirable qualities in Berkshire swine which give this breed its great economic value. At the Chicago World's Fair in 1893 Gentry's Berkshires won twenty-three of the twenty-eight first prizes offered, and many of the other winners were descended from stock bred by him. At the St. Louis World's Fair in 1903, all the Berkshires which came within the five cash prizes offered, with three exceptions, were descended from animals bred by Gentry. The entire Gentry herd was strongly in-bred, carrying a very high proportion of the blood of Longfellow. Describing his breeding practice, Gentry says:[1] "It has long been conceded that Longfellow 16,835 was the greatest boar known to the breed, in this country at least. He was out of an imported sow and one of the best I ever saw. His sire, Charmar's Duke 13,360, I bred, and he was a great one, too. He was sired by an imported boar and was out of an imported sow, and this sow and the dam of Longfellow were got by the same boar. After the death of Charmar's Duke 13,360, which happened when he was only a two-year-old, I kept his best son, Longfellow, and after Longfellow's death his best sons, and after their death, their best sons. Thus Longfellow, Longfellow's sons, and now his grandsons, have followed each other in use on my herd. One of the largest and best boars I have ever produced, one which I showed at the World's Fair at Chicago in 1893, weighing at 13 months and 6 days of age, 660 pounds, with as much action, strength, vigor and masculine development as any boar I ever saw, was produced by a son of Longfellow out of a daughter of Longfellow sired by the sire of Longfellow. Thus the three top successive sires in his pedigree were the sire of Longfellow, Longfellow and a son of Longfellow. I could name many good ones bred as closely and, in fact, almost every animal in my herd has been produced by as close in-breeding on both sides." (See Plate XV.)

"I have practiced in-breeding more from necessity than from any other reason. I believe I have not used a boar other than my own breeding for twenty years."

After a lifetime's experience with in-breeding the conclusions of Gentry are significant: "If it is true that in-breeding intensifies weakness of constitution, lack of vigor, or too great fineness of bone, as we all believe, is

[1] Gentry, Amer. Breeders' Assoc., vol. I, 1905, pp. 170–171.

it not as reasonable and as certain that you can intensify strength of constitution, heavy bone, or vigor, if you have those traits well developed in the blood of the animals you are in-breeding? I think I have continued to improve my herd, being now able to produce a larger percentage of really superior animals than at any time in the past..."

Another noted breeder of Berkshires who attributes the high quality of his animals to the practice of in-breeding is A. J. Lovejoy of Illinois. He is quoted in Davenport's "Principles of Breeding" as follows:

[1] "We are believers in quite close, even in-breeding. We find the greatest show animals are closely in-bred. Sires to half-sisters is the most common form of close breeding, though cousins, nephews, and nieces, and even brothers and sisters are bred together with great success. It of course requires good judgment in mating animals that are particularly strong in individual merit. Should each have a bad defect in any way, we should expect that to be more manifest in the offspring than in the parents, and likewise the good points would be better; so if one mates equally good specimens the produce will be an improvement. There is no sire of any breed so prepotent as an in-bred sire. When we get to the point where we feel the need of outside blood we mate an imported sow with our best boar, and from this litter we select a boar to use on the get of his own sire from other sows in the herd; that is, we breed this boar on his own half-sisters."

219. In-breeding corn. — Many of the principles which are now universally accepted as guides to practice in animal-breeding were first discovered by plant-breeders. The mendelian principle is a good example of this fact.

---

[1] Davenport, "Principles of Breeding," p. 625.

While the fundamental principles governing the transmission of characters are the same in plants and animals, the practical applications are sometimes widely different. Wholesale advice based wholly upon plant investigations, therefore, and intended to guide the practical breeder of animals is not always justified for obvious reasons. The experiments conducted for the purpose of testing the effects of in-breeding on Indian corn (*Zea Mays*) have been extensive. In every case, so far as the author has been able to determine, in-breeding corn has resulted in decreased yields if long continued. Hayes[1] and East found that "the first generation of in-breeding has the greatest detrimental effect." The injury from in-breeding is not continuous, but results in separating the pure lines in a variety. When once a pure line is produced by in-breeding, further in-breeding apparently does not reduce the yield. Shamel[2] found that four generations of self-fertilization of corn so weakened the strain that the seed failed to germinate. Darwin, experimenting with morning glories (Ipomæa) for a period of ten years, found that the strain which was screened so that insects could not bring about cross-fertilization lost in vigor as compared with a similar strain which had beeen left in the open and cross-fertilized through the agency of insects.

Wheat is a self-fertilizing plant. Unnumbered generations of in-breeding seem not to have decreased its vigor or lowered its fertility. Castle in-bred brother and sister of Drosophila (pomace fly) for 59 generations. No loss of fertility or vigor was observed.

[1] Hayes and East, Bul. 168, Connecticut Agricultural Experiment Station, p. 11.
[2] Shamel, U. S. Dept. of Agr., Year Book, 1905, p. 388.

Among the domestic animals it is quite probable that the effects of in-breeding on different species will differ materially.

220. How long is it safe to continue in-breeding. — If we limit the application of the term in-breeding to matings between parent and offspring or between brother and sister, then we cannot escape the conclusion that long-continued in-breeding results in decreasing fertility, and probably also weakens the constitution and decreases the size of the offspring.[1] "Continued in-breeding," says Kraemer, "always must result in weakened constitution through its own influence." But such results follow long-continued in-breeding. What are the limits of safety? How long may the domestic animals be closely in-bred without injury? An answer to these questions is only possible when all the conditions are known, including a knowledge of the inherited tendencies of the in-bred animals. But it seems entirely safe to conclude from the evidence available that the almost universal prejudice against the practice of in-breeding is in a large degree unwarranted. Such a prejudice has undoubtedly limited the usefulness of many valuable breeding animals and has caused real economic loss to many breeders.

221. Selection important. — The practice of in-breeding will never be successful in the absence of rigorous selection. As the undesirable qualities are transmitted with the same intensity as the good, constant vigilance is required to guard against bringing forward latent characters which are less desirable. Particularly animals

---

[1] Kraemer, "Mitteilung der Deutsches Landwirtschafts Gesellschaft," September, 1913. See also *Journal of Heredity*, 1914, p. 226.

which are undersize, weak in constitution, or that show a tendency to low fertility should not be in-bred. It is not always possible to detect the presence of these tendencies in a single generation, hence a knowledge of family history and pedigree is important. Given an animal of unusual merit with strong constitution, good size and strongly fertile, the breeder runs little risk in practicing in-breeding for a limited time.

222. The truth about in-breeding. — In the midst of such diversity of opinion as exists concerning the results and value of in-breeding, the practical breeder may well be puzzled. Sweeping generalizations either for or against the practice are apparently unwise at this time. An unreasoning prejudice against the practice will result in withholding from breeders a valuable method of breeding which has been in many cases the chief reliance of the world's greatest improvers of the domestic animals. On the other hand, a blind following of those enthusiasts who have claimed for in-breeding some mysterious power in the improvement of animal character will certainly lead to disaster.

The practice of in-breeding has been compared to a powerful medicine which in the hands of a skillful physician may decide the issues of life, but in the hands of the novice becomes a dangerous and often fatal instrument. In-breeding may be practiced successfully, but only by those who are familiar with the biological principles involved, and who are familiar with the results which sometimes follow the mating of nearly related animals and, what is quite as important, who know the ancestral history of their breeding stock.

Whether we accept the view that evil is an incidental result due to the intensification of undesirable qualities already existing in the germ-plasm of the parent stock, or whether we hold that certain definite evils are a necessary result of mating animals of near kin, we must admit that in-breeding has often been practiced with great success and no appreciable injury. It is, therefore, clearly apparent that there are conditions which are neither unusual nor extremely rare under which in-breeding can be practiced with the assurance of success.

223. Fixing characters by in-breeding. — In-breeding has been a powerful means of fixing and perpetuating valuable characteristics in the domestic animals. It is still a valuable method to be employed for the same purpose. But in-breeding is only a means to an end and not the end. In-breeding possesses no magic or occult power which will be exerted for the improvement of animals. While it works powerfully in fixing the good qualities, it is no less potent in firmly establishing undesirable qualities which may be present in the parent stock. And in this fact lies the chief danger from in-breeding. In fixing the good characters, we may unconsciously strengthen the powers of transmission in the direction of bad qualities. The most skillful breeders are less likely to err in the direction of perpetuating tendencies to evil, and history gives ample confirmation of the certain good which does follow in-breeding when practiced with intelligence by skillful and experienced breeders and accompanied by rigorous selection.

It has often happened in the experience of breeders that a sudden mutation has appeared in a single animal. This mutation may represent a high degree of improvement in a certain character or characters which the breeder has long sought to develop in his breeding stock. This variation appears in one animal only. It is highly desirable not only to perpetuate this improved character but to breed animals that have this quality in as pure and dominant a state as in the original animal. This can be accomplished by in-breeding. No other method is available which will so quickly and certainly result in producing offspring of similar or identical blood lines. The history of animal-breeding is rich in instances of great animals, famous for their individual excellence but more famous because they have left a heritage of potent " blood " which has established a new and better strain or even a new breed. We need only recall the names of Favourite, the Shorthorn bull, Justin Morgan, the founder of the Morgan breed, Hambletonian 10, the forerunner of the American trotting horse, and scores of individuals of lesser note belonging to practically every modern breed. In many of the instances to which reference is here made, the great individuals would never have become famous if breeders had not recognized their peculiar excellences and have insured the perpetuation of their valuable characters by in-breeding.

224. In-breeding and prepotency. — The prepotency of animals is increased by in-breeding. There is unanimity among investigators and practical breeders on this point. By continuous in-breeding we may " breed out " less desirable qualities, that is, in the light of mendelism the characters in the germ-plasm tend to become homozygous. In-bred animals are " pure-bred " animals not only in the parlance of the breeder but also from the standpoint of genetics. Mating animals of diverse characters tends to destroy prepotency. Mixing the blood of animals of widely differing characteristics results in making the constituent characters of the germ-plasm heterozygous. The cross-bred animal is never prepotent.

225. Results of in-breeding vary with different species. — The varying opinions regarding the benefits or injuries from in-breeding may in part be accounted for from the fact that investigators have based their conclusions upon data gathered from researches on widely differing species of animals and plants. The effects of in-breeding are quite different in different species or families. Among plants, nature seems to have designed some species especially to insure cross-fertilization and to guard against self-fertilization, while other species are self-fertilizing. Indian corn (*Zea Mays*) is a good example of the former class of plants. The results of continuous in-breeding on the maize plant are markedly injurious. Shull found that continual self-fertilization in Indian corn resulted in a loss of vigor. There are other plants like wheat that are self-fertilizing, and it is difficult to see how in-breeding can be injurious in such species.

CHAPTER XII

CROSS-BREEDING

THE term crossing or cross-breeding, like the term in-breeding, is not capable at this time of exact definition. In general we may define cross-breeding as the mating of individuals which are not related. The literature of the subject indicates that this term has been loosely applied. Some indeed have used the term to designate the mating of individuals belonging to different families within the same breed. As a rule, cross-breeding means the mating of individuals belonging to different breeds, as a cross between the Shorthorn and Hereford; or the union of animals belonging to different species, as a cross between the stallion and the jennet. Cross-breeding has been strongly recommended by some breeders as a valuable method of improving the domestic animals.

226. Permanent and temporary results of cross-breeding. — In recommending cross-breeding, the advocates of this practice have not always been careful clearly to differentiate between the permanent and lasting results of cross-breeding and the more immediate and temporary advantages. The effect of cross-breeding upon the purity of the heritable characters of the breed as represented by the germinal elements in the germ-plasm is one thing, while the more or less temporary effect on the body-cells may be quite another thing. The purpose of the breeder of pure-bred registered animals is to establish a race of animals that will breed true. It matters not how many good qualities the individual breeding animal may possess; if he cannot transmit these good qualities to his offspring, he is not a desirable animal for breeding purposes. He may be valuable for commercial purposes. Such an animal might be a fast horse, a prize-winning beef animal, or a great producing cow, but lacking the ability to transmit these qualities this animal would not be a desirable or valuable individual in a breeding herd. It may be quite possible, therefore, for a method of breeding to have a distinct economic value for the production of commercial animals and at the same time be a very bad method for the breeder of improved live-stock whose purpose is to produce animals for breeding purposes and not for slaughter or work. What effect does cross-breeding have on the breeding powers of the domestic animals? What value, if any, has cross-breeding in the production of animals for commercial purposes and which are not intended to be used for breeding? The breeder's interest in cross-breeding will naturally center about the relations of this practice to heredity.

227. Advantages from cross-breeding. — Breeders of the domestic animals have frequently practiced cross-breeding in the belief that certain very definite and specific benefits followed such practice. In attempting to analyze the reasons for practicing cross-breeding, it is apparent that this has been generally followed for one or more of the following reasons, — to increase fertility, to restore weakened constitution, to increase the size or for improvement.

228. Grading. — The practice of grading, by which is meant the improvement of native or unimproved animals by mating with pure-bred or registered animals, should not be confused with cross-breeding. Cross-breeding is the union of two or more distinct races or breeds, while grading is an attempt gradually to develop a type by continually breeding to pure-bred sires.

Grading is one of the most successful and certain methods of improvement. There are many examples of successful grading among the breeders of domestic animals. Manifestly the more inferior the foundation mother stock, the greater will be the improvement when mated to a pure-bred registered sire. A few generations will often suffice to produce " high-grade " cattle, horses, sheep or swine that will possess most of the valuable qualities which have commercial value. For commercial or economic purposes, the high-grade beef animal may be as valuable as the pure-bred. A high-grade dairy cow will often produce as much milk and butter as the registered cow. For breeding purposes, the pure-bred registered animal is far superior. The grade does not transmit its qualities with certainty. One object of pure breeding is to develop the quality of prepotency, and this is accomplished by long years of most careful selection and mating. The grade animal cannot possibly possess the quality of prepotency to the same extent as the pure-bred form, hence it follows that even if the grade does exhibit a high degree of individual merit, this is no evidence of ability to transmit the same qualities to the offspring. A high degree of individual excellence in a pure-bred registered animal is more certain to be transmitted, and for this reason the registered animal of high merit is often held by the experienced breeder at values which seem beyond any real economic basis.

229. Cross-breeding to increase fertility. — Some animals are infertile when bred to other individuals of their own breed. This is particularly the case if the two animals mated are the result of long-continued in-breeding and are themselves also near of kin.

In the instance of in-breeding pigs by J. Wright already cited, the seventh generation resulting from close in-breeding consisted of one sow. This sow was infertile when bred to her sire, but bred readily with an unrelated boar. Darwin cites numerous instances of increased fertility due to crossing. Mr. Eyton,[1] a breeder of Grey Dorkings, found it necessary to increase the prolificacy and increase the size of his in-bred stock by crossing. Bates,[2] the great breeder of Shorthorn cattle, bred closely in-and-in for thirteen years, but then found it necessary to "infuse fresh blood, not to improve the form of the animals but on account of lessened fecundity."

Bloodhounds[3] closely in-bred lost their fertility, which was restored by a single cross.

Many plants are infertile unless cross-fertilized with the pollen of another variety.

230. Cross-breeding to increase size and restore constitution. — The tendency of in-breeding to decrease the size is promptly corrected by cross-breeding. "The good effects of a cross are at once shown by the greater size of the offspring."

It is the common experience of breeders that highly improved strains of cattle, hogs or sheep sometimes show a refinement or delicacy of constitution which in a measure interferes with the economic value. In such cases a sudden out-cross to another equally valuable strain may

---

[1] Darwin, "Animals and Plants under Domestication," vol. II, p. 105.
[2] *Ibid.*
[3] *Loc. cit.*

quickly correct any tendency to inferior size or weakened constitution. Not only are in-bred animals benefited in certain definite qualities by crossing, but breeds and families which have not suffered in any way from in-breeding are sometimes improved in size, vigor and fertility by crossing.

231. Crossing and heredity. — As in-breeding tends to simplify the germ-plasm and strengthen the powers of transmission, so cross-breeding tends to weaken the prepotency and complicate the elemental constitution of the hereditary substance. Crossing has a tendency to break up established characters. It destroys combinations of characters which have long existed in the strain and which under systems of pure breeding have behaved in a manner like unit characters in transmission. The result of crossing pure-bred animals is often to destroy the results of generations of careful breeding and selection.

232. First cross an improvement. — The cross-bred offspring of pure-bred parents often show an improvement over either of the parents. This superiority may be exhibited not alone in increased fertility and more vigorous constitution, but also in the very qualities which characterize the parents. A cross between animals belonging to distinct breeds may be a better beef animal than either parent. The Scotch farmer breeds the Aberdeen Angus cow to a white Shorthorn bull. The offspring is the well known " blue gray " which is highly prized by the feeder and in the fat cattle market commands a premium. over the pure-bred animals of either breed. The fat cattle exhibitions of the world have not infrequently given the highest prizes of the show to cross-bred animals. (See Plate XVI.)

But when it is attempted to perpetuate the superior qualities of the cross-bred animal by breeding, disappointment invariably results. The second cross resulting from the mating of two cross-bred animals may be totally unlike either of the immediate parents or of the original pure-bred forms. Crossing, therefore, is not a method to be employed for rapid improvement or for fixing desirable qualities. It is opposed to in-breeding which does increase prepotency and is the most rapid method known of fixing desirable characters.

233. Cross-breeding as a cause of variation. — The fact that crossing disturbs the balance of characters and brings about recombinations in the germ-plasm gives it a peculiar value in causing variations to appear. The breeder who is working with pure-bred animals which owe their purity of breeding to a long period of careful selection by skillful breeders cannot hope to cause any great degree of improvement. Pure-bred animals are already improved. About all any breeder working with pure-bred animals can do is to select out the highly desirable strains from those of lesser value already in the breed. But as Johanssen has shown, there are very definite limits beyond which the improvement of pure lines cannot go. Marked improvement must come through variation. Crossing is a common cause of variation. Variations which appear as the result of crossing may be desirable or undesirable. They may be relatively unimportant or they may be in the nature of a valuable mutation. Such valuable mutations may be perpetuated by in-breeding and a new and valuable quality sometimes secured in this way. This method is not practical for breeders of registered live-stock under present conditions, but has been of great service to the breeders of plants.

234. Crossing species. — Many species may be successfully crossed. Some of these crosses are of great economic value, as the cross between the mare and the jack. The number of successful crosses between animal species is not large. Such unions are difficult to make and generally sterile. When such crosses are possible and the union is fertile, the offspring is generally partially or wholly sterile. Some of the successful crosses which have been reported are the sheep and goat, horse and ass, horse and zebra, cattle and yak, cattle and bison, brahmin and domestic cow, game cock and guinea fowl, domestic fowl and pheasant, dog and wolf, and dog and fox.

235. Crossing bison and cattle. — A most interesting experiment in cross-breeding between the bison and domestic cattle is reported by Mossom M. Boyd.[1] The hybrid offspring from Hereford dams and bison sire were very uniform, all having white faces, were larger than the bison and much smoother, broader and deeper than the sire. Great difficulty was experienced in making the first cross from the excessive secretion of the amniotic fluid. This difficulty caused many deaths. The percentage of males from the first cross was very small. Among forty-five hybrids, only six were males. Of these three died at birth, one died in less than twenty-four hours after birth, one proved barren, and the last male was killed before determining his fertility. Charles Goodnight of Texas reports[2] that "no male calves have ever been born; cows conceiving them either suffer abortion or die, hence only get heifer calves and only a small per cent of them." The hybrids produced their first calves at an average age of five years (Plate XVII).

[1] Boyd, *Journal of Heredity*, 1914, p. 189.
[2] Goodnight, *Journal of Heredity*, 1914, p. 199.

Three-quarter blood bisons from pure bisons, bulls and hybrid cows were similar in form and color to the bison; one cross-bred from a half Hereford dam had a white face. One-quarter blood bisons from hybrid dams and Hereford and Aberdeen Angus bulls were uniform in conformation but varied in color. The three-quarter bloods closely resembled the bison, while the one-quarter bloods could not readily be distinguished from domestic cattle. From twenty-four hybrid cows, only three were regular breeders and fifteen were barren. From twelve one-quarter blood bison cows bred to domestic animals, seven were fully fertile, four were irregular breeders, and one was barren. One out of four of the three-quarter blood bison cows was barren. The term "cattalo" is used by Boyd to designate the third generation. When both parents are of mixed blood, the cattaloes are in many respects superior to ordinary domestic cattle, being hardier and much less subject to disease. Cattaloes with a high percentage of bison blood are probably immune from Texas fever and blackleg. The cattalo grows to a greater weight than domestic cattle. Goodnight[1] says, "More of them can be grazed on a given area. They do not run from Heel Flies nor drift in storms. They rise on their fore feet instead of their hind feet. They never lie down with their backs down hill, so they are able to rise quickly and easily."

It seems entirely probable that a new breed will be added to the list of domestic cattle, and if this result is achieved, it will be one of the very few authentic cases of the establishment of a new breed by crossing species.

236. The mule hybrid. — The most widely distributed and most useful hybrid known is the mule, which is pro-

---

[1] Goodnight, *Journal of Heredity*, 1914, p. 199.

duced by crossing the domestic mare to the jack. In 1915 there were 4,479,000 mules in the United States. This was more than one-fifth of the total number of horses in the country at the same time. The production of mules has increased at a more rapid rate than horses, and the use of mules is becoming more extensive. The mule hybrid is a remarkable example of the practical advantages which follow a particular cross. This animal is more hardy and enduring than either parent. As compared with the horse, the mule is longer-lived, less subject to disease or injury, and more efficient in the use of food. The mule can be safely put to work at a younger age, will thrive on coarser feed, and seems to be much better able to avoid many dangers which menace the usefulness of the horse. The mule will perform more arduous labor on less food. The mule will endure the heat of southern latitudes more successfully than the horse and is therefore a popular draft animal in the South.

The cross between the mare and jack is readily accomplished and the union is perfectly fertile. The conformation of the mule more closely resembles that of his sire. The ears are long, feet long and narrow, withers sharp, mane and tail scanty, and the voice a bray like the jack. The mule is sterile. A few cases of supposed fertility of mare mules have been reported, but the writer has investigated several apparently reliable reports and has never found an authentic case of a fertile mule. Most of the erroneous reports of fertile mules have apparently arisen from the not infrequent cases of mare mules which have been observed suckling mule foals. The milk glands of mare mules have been known to function as the result of the stimulation afforded by a suckling foal. A case of a mare mule giving milk was reported to the writer by L. O. Swarner of Boonville, Missouri, in 1913. This mare was six years old and at the time had been giving milk for five weeks. The milk glands had not been stimulated in any way, but the milk "streamed" from the udder. It is also of interest to know that this mare mule showed unmistakable evidences of what in the ordinary mare would be regarded as complete sexuality. She came in heat regularly. The mule was bred frequently when in season to both the stallion and jack, but failed to conceive. A sample of the milk was analyzed by the Chemical Department and found to contain 2.46 per cent protein, 5.8 per cent sugar, 1.45 per cent fat, and .4 per cent ash. (See Plate III.)

The mare mules apparently have all the essential organs of reproduction and come in heat with considerable regularity. The horse mule also has the essential sexual organs well developed and his sexual instincts are so well developed that castration of young mules is universally practiced. The cause of sterility in the horse mule is not due to a failure to develop spermatozoa, but the sperm-cells are imperfect. In some cases the sperm-cells lack the tail or flagellum.

237. The hinny hybrid. — The reciprocal cross between the jennet and the stallion is accomplished without difficulty and the union is very fertile. The hybrid from the cross is called a hinny. Some authorities have held that the hinny resembled the horse much more closely than the mule, but this is denied by most practical breeders. The hinny is not commercially important as the jennet is too valuable for the production of jacks to be used for crossing. The hinny is sterile. (See Plate XVIII, upper.)

238. Crossing the horse and the zebra. — The horse and zebra have been successfully crossed by Ewart of

Edinburgh, Scotland, the United States Department of Agriculture and many others. The zebra-horse hybrid is easily domesticated and can be successfully broken to harness. The first cross is not so easily made as that between the jack and the mare but is not impossible or extremely difficult. The zebra possesses a much smoother, finer and more horse-like form than the ass, and the zebra hybrid therefore is possessed of more quality and "finish" than the mule. This hybrid should prove valuable, particularly in those regions where the "tsetse" fly is fatal to horses but not to zebras and probably not to the zebra hybrids.

239. Crossing cattle and zebu. — Many crosses have been made between the zebu and European cattle and between the zebu and the cattle of Tunisia. The first cross in practically all of the experiments seems to have been successful.

The cross-bred zebu is resistant to Texas fever and anthrax and is not seriously inconvenienced by foot and mouth disease.[1] In Brazil [2] the zebu cross is popular. It is claimed that the cross-bred zebu is more prolific and that these animals herd together better than the ordinary domestic cattle. The zebu hybrids are less tractable and docile than domestic cattle, but are very active and enduring draft animals.

Because of the disease-resisting qualities of the zebu, its prolificacy, adaptability to hot climates and general hardiness, Nabours [3] is of the opinion that this type of cattle may yet become an important breed in the United States.

[1] Roederer, *Journal of Heredity*, 1915, p. 201.
[2] Hunnicutt, *Journal of Heredity*, 1915, p. 195.
[3] Nabours, *American Breeder's Magazine*, 1913, p. 38.

240. Sheep-goat hybrid. — The cross between the sheep and goat has been successful in a number of instances. Spillman reports such a hybrid belonging to E. Armand of Monett, Missouri. (See Plate XVIII, lower.)

The covering of the body of this hybrid was generally goat hair, but the back was covered with " shaggy wool." " This hybrid is a female and appears to be infertile, but not absolutely so, for it has once produced a half-grown fœtus." [1]

[1] Spillman, *Journal of Heredity*, 1913, p. 69.

CHAPTER XIII

DEVELOPMENT

THE qualities which an animal possesses are due in the first place to inheritance and in the second place to the manner in which the inherited qualities have been developed. An animal cannot develop beyond the capacities which have come to it through the germ-plasm. It is also true that the capacities which are inherited cannot benefit the individual unless they are developed through a favorable environment. It is seldom that an animal realizes fully the possibilities for development which are inherent in the germ-plasm. The carefully bred beef animal inheriting those valuable qualities of early maturity, broad, deep and rounded form, rugged constitution and quiet temperament, with a distinct tendency to lay on fat when food is abundant, may completely fail to exhibit these inborn characters and actually display the form and characteristics of the unimproved animal if it has been surrounded by conditions which are unfavorable for the development of these special qualities. The highly improved dairy cow with the inherited capacity to produce enormous quantities of milk and butter may never rise above mediocrity if she is not supplied with food and her milking functions intelligently developed. In the selection of animals for improvement, the skillful breeder can never know what results he has achieved until the products of his skill have been fully developed.

It is not too much to say that no man can be a successful breeder who is not also skillful in developing his animals. Thus, in practice, development becomes supremely important and throughout the history of animal-breeding has been only second in importance to heredity itself. A satisfactory treatment of development in all its phases as related to animal husbandry would require a volume, and as the chief purpose of this work is to consider how the inherited capacities of animals finally appear as definite characters in the germ-plasm, only a limited reference can be made to developmental phases of chief importance to the animal-breeder.

241. Growth. — From the fertilization of the egg until the full development of the mature individual, the animal increases in volume and changes in form. This increase and change of form is called growth. The final size of an animal is determined by the rate of growth and the length of the growth period. The guinea pig and rabbit come to full maturity at about the same age, but the rabbit is larger because its rate of growth is more rapid. The rate of growth in the rabbit and man is about equal, but man is much larger at maturity because the period of growth is much longer.[1]

242. The growth impulse. — The young of any species tend to develop and grow in accordance with the normal habit of the species. This applies in a special sense to the skeletal system. Even in the absence of a sufficient supply of feed and other favorable conditions, the young animal displays a remarkable physiological impulse to continue to increase in the skeletal parts. [2] Animals fed

---

[1] Morgan, "Experimental Zoölogy," p. 245.

[2] Waters, "Capacity of Animals to Grow under Adverse Conditions."

on a limited ration will continue to increase in height, length of body, and other parts of the skeleton, at the same time becoming thinner and thinner in flesh. Even during starvation the same tendency is apparent. This fact has been noted by H. Aron,[1] who found that while fasting, the skeleton grows at the expense of the other body tissues.

243. Factors influencing growth. — The chief factors influencing growth in the domestic animals are food, heat, light, age, gestation and lactation. The chief condition influencing growth in normal animals is the food supply.

244. Growth and food supply. — While it is true that the animal may for a limited time add to its tissues when food is insufficient in either quantity or quality, it is also true that a long-continued deficiency in the food supply of young growing animals will invariably check their growth. The check to growth in such cases may be only temporary, or it may result in permanently decreasing the normal size of the mature animal. (See Plates XIX and XX.)

245. Capacity to grow. — The young animal that is stunted as a result of insufficient food does not lose the capacity to grow. The organism seems to be able to continue to function and maintain a certain equilibrium. If later a greater abundance of food is supplied, the rate of growth may be reëstablished. If after a period of partial starvation the food supply is abundant, the rate of growth may for a time be even more rapid than before. That the capacity of an animal to grow is not destroyed by stunting is shown by the results of an investigation at the Missouri Experiment Station by Waters and

[1] H. Aron, Exp. Sta. Record, vol. 24, p. 765.

P. F. Trowbridge.[1] These investigators fed a beef steer from the age of three months to thirty-eight months old. From three to twelve months of age the animal was fed on a maintenance ration (Plate XXI). An attempt was made to feed the young calf in such a way that it would neither gain nor lose in live weight. At the beginning of the period (three months old) (Plate XXI, upper left) the animal weighed 175 pounds; at twelve months (Plate XXI, upper right) the animal weighed 212 pounds. Another animal similar in every way at the beginning was fed a full ration of nutritious feed. The latter animal increased in weight from 200 pounds at four months (see Plate IX) to 875 pounds at 336 days of age (Plate VIII, upper). The animal fed on a sparse ration continued to increase in height, length of body, size of bone, and other skeletal measurements but lost constantly in fat and gradually became leaner and thinner. At the end of the twelve months the steer was much emaciated and showed symptoms of starvation. From the standpoint of the practical feeder, the animal was clearly stunted in its growth, and in the opinion of many breeders he had lost very greatly in his capacity to gain in live weight and to do so on what would be regarded as a normal amount of food. In other words, his economic value for the production of beef was very greatly diminished. After twelve months the animal was given a gradually increasing amount of nutritious food until he was consuming a normal ration. The animal rapidly improved in condition and at twenty-four months (Plate XXII) had reached a total weight of 1055 pounds, a total gain of 842 pounds in twelve months. This gain was not expensive in that a large amount of feed was required to

[1] Missouri Experiment Station, unpublished data.

produce a pound of gain, but on the other hand the gain in live weight was accomplished by feeding only five and six one-hundredths pounds of grain and two and four-tenths pounds of hay for each pound of gain. The steer that had been fed generously for the first twelve months of its life gained only 500 pounds during the period in which the stunted one had gained 842 pounds. In the production of the 500 pounds the full-fed steer had consumed nine and eight-tenths pounds of grain and four and two-tenths pounds of hay for each pound of gain in live weight. Over forty per cent less feed was required by the stunted animal for each pound of increase in live weight. Not only did the stunted animal not lose its capacity to grow, but in certain respects its growth processes were accelerated during the period covered by this experiment as a result of its difficult struggle for existence during the first twelve months of its life.

246. Growth and the cell. — Increase in the size of animals which follows growth is due to a multiplication of cells and not to an increase in their size. The size of cells varies between rather narrow limits. The larger size of some animals is not due to larger cells in their organization but to a larger number of cells. It is also true that the cells in any individual animal vary but little in size. The increase in size of any part of an animal is due, therefore, to an increase in the number of cells and not to an expansion of cells already formed. It is true, of course, that certain minor exceptions to this rule are to be observed. The nerve cells vary in size with the size of the animal. The nerve cells of an ox are much larger than those of the pig. The frog has very large cells, while the starfish is composed of small cells. But a large frog differs from a small frog in the number of cells, not in their size. Each individual animal begins its existence as a single cell. Through cell division the embryonic organism rapidly increases in size until maturity is reached. The rate of growth of the individual is most rapid during the very early stages in the development of the embryo. The rate of growth decreases gradually from this time until full maturity, when nominally growth ceases.

247. When the growth impulse is strongest. — The growth impulse is strongest in the animal while still existing in the uterus of the mother. After birth the growth continues less rapidly, but is still very rapid when compared with the increase in size during the later months of the growth period. It is for this reason that the period of gestation is so fundamentally important in the life of the animal. During this period of exceedingly rapid increase in size and development of the vital organs and other parts of the body, any abnormal condition which interferes with the normal requirements of the unborn animal may cause arrested development and result in seriously retarding the growth or permanently crippling the individual. It is undoubtedly true that during this period the practical breeder may through skillful feeding and care materially influence for good or evil the development of the valuable characters of the domestic animals.

248. Development of the foetus. — The development of the fertilized egg through the embryonic stages of the life of the mammalian animal is influenced by a number of conditions which may have a profound influence upon the material well-being of the future mature animal. Some of these influences are as yet obscure and not well understood, while others are more clearly determined and their effects more easily recognized.

The development of the fœtus is influenced by heredity, and the physiological environment of the pregnant mother. Among the latter are the general health or well-being of the mother, age, the quantity and quality of the food supply and mental impressions.

249. Heredity and fœtal development. — The inherited tendencies of an animal are exhibited from the very beginnings of its existence in the fertilized egg-cell. Its fœtal development, therefore, must be influenced to a certain extent by those inherent determiners which have come to the fertilized egg-cell from the male parent. The size of the fœtus at various stages of development then would be determined, not alone by the maternal heredity and environment, but also by the inherited characteristics which have been acquired through the male. At the same time it is probable that the maternal environment has a much more important influence on the fœtal development than its paternal hereditary tendencies. If it were otherwise, serious consequences might follow the mating of smaller females to much larger males among the domestic animals.

Pony mares weighing 700 to 900 pounds are not infrequently mated with large stallions of the draft breeds weighing 2000 pounds or more. Under these conditions the fœtus is not as much larger than the normal fœtus of the mother as might be expected from the much greater size of the stallion. Small burro mares weighing 400 or 500 pounds have been artificially inseminated with the sperm of Percheron and other heavy draft stallions. The growth of the fœtus in this case is undoubtedly somewhat greater during the normal period of gestation than the growth of a pure burro fœtus, but the increased size is not a mean between the normal size of a Percheron and a burro fœtus, but is much smaller.

RELATION OF WEIGHT OF DAM TO BIRTH WEIGHT OF LAMB

Weight of Dams	Number of Single Lambs	Average Birth Weight of Single Lambs	Number of Twin Lambs	Average Birth Weight of Twin Lambs	Average Birth Weight of All Lambs
Below 90 lbs.	8	7.2 lbs.	—	—	7.2 lbs.
90 to 100 lbs.	6	7.4 lbs.	—	—	7.4 lbs.
100 to 110 lbs.	14	8.6 lbs.	8	6.4 lbs.	7.5 lbs.
110 to 120 lbs.	12	8.7 lbs.	20	7.2 lbs.	7.9 lbs.
120 to 130 lbs.	13	8.9 lbs.	10	7.6 lbs.	8.3 lbs.

250. Birth weight of lambs. — The author,[1] investigating the birth weight of lambs from grade Merino ewes bred successfully to Shropshire, Hampshire and Merino rams, found that the size of lambs at birth is primarily determined by the nutrition of the fœtus while carried in the uterus of the mother. The nutrition of the fœtus will of course be determined entirely by the physiological condition of the mother during gestation. This is shown in one test in which the birth weight of twenty-nine lambs sired by two heavy rams averaging 237 pounds in weight was 8.16 pounds. The average weight of two other rams of the same breeds was 142½ pounds. The two lighter rams sired twenty-five lambs from the same ewes, or ewes of the same type. The average birth weight of these lambs was 8.75 pounds. The weight of the sires in this case seemed to have little influence on the weight of lambs at birth. In attempting to discover the real factors determining the growth of the fœtus as measured by the weight of lambs at birth, it was found that the

[1] F. B. Mumford, "Some Facts Influencing the Weight of Lambs at Birth," Bulletin 53, Missouri Experiment Station.

nutritive condition and weight of the mothers had an important influence on the development of the fœtus. This fact is shown in the table on page 262.

Summarizing the results of four years' work, the author says,[1] "We must conclude from the exhibit here made, comprising the results of 61 births, that the weight of the mother has a direct influence upon the birth weight of the offspring and that in general the lambs having a heavier weight at birth are produced from the larger ewes."

251. Effect of protein and ash in ration on fœtal development. — The development of the fœtus may be materially influenced by the ration given to the mother during pregnancy. Evvard[1] has shown that if pregnant sows are fed a ration poor in protein, the pigs are smaller at birth. Not only were the offspring smaller at birth, but they were weaker and the death rate greater. The first investigation was made with young sows in 1910–1911. These were fed the rations indicated in the table on page 264 for a considerable period. The results are summarized in the table.

[2] "The basal ration was corn alone. Corn we know is quite deficient in protein (the zein which comprises practically 58 per cent of said protein is peculiarly lacking in two quite important amino acids, namely, tryptophane and lysine) and calcium. It is somewhat surprising to know that calcium comprises practically two-thirds as much of the body substance as does nitrogen, the basal element of protein."

"Note that the supplemented rations not only produced larger but stronger pigs at birth. A studied survey

[1] Mumford, *loc. cit.*, p. 176.
[2] Evvard, Iowa Academy of Science, Report 1913, p. 326.

of the above figures shows most clearly that even though the carbohydrates were limited, as in the meat meal lots, the increase in protein and ash was such as to markedly influence the size and strength of the new-born pigs. That clover and alfalfa should also have a marked effect is logical because these hays are leguminous in character, run high in protein and calcium, and also have an alkaline ash which is probably beneficial."

EFFECT ON OFFSPRING OF FEED FED PREGNANT SWINE [1] GILTS. — FIVE IN A LOT. 1910–1911

	GILT RECORD			OFFSPRING RECORD					
		Feed Daily				Per cent Vigor			
PREGNANCY RATION OF GILTS	Av. Daily Gain Lbs.	Shelled Corn Lbs.	Supplement Lbs.	Av. No. Pigs in Litter	Av. Wt. Newborn Pigs Lbs.	Strong	Medium	Weak	Dead
Corn Only	.354	3.65	None	7.6	1.74	68	16	16	0
Corn+Meat Meal (Light)	.582	3.21	0.127	7.4	2.01	92	5	3	0
Corn+Meat Meal (Heavy)	.635	2.75	0.432	8.8	2.23	93	5	2	0
Corn+Clover	.528	3.67	0.302	6.4	2.21	94	0	6	0
Corn+Alfalfa	.627	3.74	1.106	7.6	2.29	89	8	0	3

A similar test with older (yearling) sows confirmed the conclusions from the first investigation. In the latter trial there was evidence that animal protein supplied in meat meal was more efficient than vegetable protein supplied in linseed oil meal.

252. High calcium rations for pregnant swine. — The ordinary rations fed to farm animals in most localities

[1] Evvard, Iowa Academy of Science, Report 1913, p. 326.

are known to be deficient in calcium. This is particularly true when, as in the case of swine, the rations are chiefly composed of grain with a relatively small proportion of roughage. It is a popular notion that feeding pregnant farm animals a high calcium ration will cause the skeletal system of the fœtus to develop even beyond what may be regarded as normal. As a result of this greater development of the skeleton, it has been alleged that the mother may sometimes have serious difficulty in expelling the fœtus. Some confirmation of this belief is apparently found in the work of Evvard already described. But in this investigator's trials a ration abnormally deficient in calcium and protein is compared with normal rations. Adding an excess of calcium to a normal ration may not necessarily increase the size or change the composition of the fœtus. In this respect we know that similar changes in the ration do not change the composition of the milk. At the Wisconsin Experiment Station,[1] Hart, Steenbock and Fuller compared normal rations with a high calcium ration fed to pregnant swine. Eight Poland China sows were fed in lots of two each. Three lots were fed on a high calcium ration by the addition of calcium carbonate, calcium phosphate (floats) and alfalfa. One lot was fed a normal standard ration known to be adequate for pregnant swine. These rations were fed throughout the gestation period. Although the calcium rations contained five times the amount of this mineral present in the normal ration, there was no evidence that the skeleton of the fœtus was influenced in any degree. The authors in summarizing their results

[1] Hart, Steenbock and Fuller, "Calcium and Phosphorus Supply of Farm Feeds and their Relation to the Animal's Requirements," Wisconsin Research Bulletin No. 30.

conclude that "High calcium rations, as compared with low calcium rations, had no effect whatever during a single gestation period on the size or calcium content of the skeleton of the fœtus. The skeleton is not increased in any dimension by a wide variation in the amount of calcium fed the mother."

253. Size and vigor of fœtus as influenced by corn and wheat rations. — The particular rations fed to breeding animals may profoundly influence the character of the fœtus. It is not enough that the prospective mother should receive a ration containing the right proportions of protein, carbohydrate and mineral substances, but these minerals must be of the kind of substances which are known to satisfy the nutritive necessities of the animal. A most interesting demonstration of this fact was made possible through the valuable work of Hart,[1] McCollum, Steenbock and Humphrey at the Wisconsin Experiment Station. In May, 1907, these investigators began feeding four heifers a ration of corn meal, corn stover and gluten feed, all corn products. Another group of four was fed wheat meal, wheat straw and wheat gluten, nutrients derived entirely from the wheat plant. The materials supplied in each ration were proportioned in such a manner as to furnish each group of animals a well-balanced ration in accordance with the accepted feeding standards. The feeding continued for two years and accurate records were kept of feed consumed, gains in live weight, and physiological condition of all animals in the experiment.

[1] Hart, McCollum, Steenbock and Humphrey, "Physiological Effect on Growth and Reproduction of Rations Balanced from Restricted Sources," Research Bulletin No. 17; Wisconsin Experiment Station.

The rate of growth of all animals in the two groups was very similar, indicating little difference in the efficiency of the rations. But that there was a difference is indicated by the author's description:[1] "The corn-fed animals (Plate XXIII) looked smooth of coat, fuller through the barrel; and, as expressed by experienced feeders and judges of domestic animals, they were in a better state of nutrition. On the other extreme stood the wheat-fed group with rough coats, gaunt and thin in appearance, small of girth and barrel, and to the practical eye, in rather a lower state of nutrition." But perhaps the most significant results of this experiment were the effects of these rations on the reproductive functions of the mothers and the vitality of the offspring. The gestation period in the corn-fed mothers was practically normal, while in every case the calves of the wheat-fed group were dropped from two to five weeks before the end of the normal gestation period. The calves from the corn-fed cows were uniformly strong and vigorous and were normal in every respect. The calves from the wheat-fed cows were all born dead or with such low vitality that they soon died.

The yield of milk from the wheat-fed (Plate XXIV) cows was distinctly below that of the corn-fed group. In discussing the significant results of this investigation the authors say, "These results emphatically show how depressing or stimulating the influence of a ration may become, even when it is made up of supposedly normal feed materials and balanced as to ordinary chemical constituents and supply of energy, especially when that ration is continued for a long time. The evidence for the necessity of giving much weight to the physiological influence of the ration,

[1] *Loc. cit.*

apart from its digestible protein content and calorific value, is positive."

254. The permanent effect of retarded growth. — Is the retardation of growth resulting from unfavorable environment a permanent condition? Is it possible permanently to stunt an animal? This question is one of great interest to the practical breeder of live-stock and the live-stock farmer. The animal is often employed by the farmer for the purpose of disposing profitably of food materials which are of limited value from the standpoint of nutrients and digestibility, but nevertheless have a food value. Teachers and investigators in animal husbandry have for a long time taught that any condition which resulted in stunting the young animal, permanently affected the mature individual. It has also been claimed by some that the capacity to grow was materially diminished by a stunting period.

The history of two animals which were fed at the Missouri Experiment Station by P. F. Trowbridge[1] is of great interest in attempting to answer this question. One of these animals was given a full ration from birth and the other animal was given a so-called maintenance ration from three months of age to thirteen months. The use of the term maintenance ration in this connection means that it was planned to feed the animal sufficient food to cause it to maintain a uniform body weight. As a matter of fact, the animal added somewhat to his total weight in the ten months of stunting. The tables and pictures (Plate XXV) give a very clear idea of the general results of this trial. The full-fed animal gained rapidly and at the end of fourteen months weighed 956 pounds. The young animal fed merely a maintenance ration weighed

[1] Missouri Experiment Station, unpublished data.

only 207 pounds at the age of thirteen months. At that time, the animal was emaciated in appearance, showed every symptom of starvation, and was to all appearances very severely stunted in its growth and development. From twelve months to thirty-eight months, the animal fed previously on a maintenance ration was later given a generous ration and gained during that period a total of 1280 pounds. The full-fed animal during the same period gained 853 pounds. At the end of the period the animal that was stunted during its early life was over 300 pounds lighter, was apparently somewhat shorter, with finer bones, than the full-fed animal. There was little doubt but that in this trial the early environment permanently decreased the size of the adult animal.

255. Early stunting and the capacity to grow. — It is interesting to know that the capacity to grow was not

The Permanent Effects of Retarded Growth

Feed and Weight Records of Two Steers for 24 Months

No. 527 was fed a generous ration until 38 months of age. No. 529 was starved for the first 12 months and then given a full ration until 38 months of age.

	Age 1 to 12 Mos.		Age 12 to 18 Mos.		Age 12 to 24 Mos.	
	No. 527	No. 529	No. 527	No. 529	No. 527	No. 529
Initial Weight (lbs.)	165.0	175.0	901.8	213.5	901.8	213.5
Final Weight (lbs.)	901.8	213.5	1178.2	694.9	1401.7	1054.7
Gain during Period	736.8	38.5	276.4	481.40	499.9	841.2
Av. Daily Gain	2.046	0.106	1.535	2.674	1.388	2.336
Grain Fed Daily	6.926	0.571	14.686	10.030	13.610	11.829
Hay Fed Daily	3.578	0.895	6.401	4.840	5.821	5.642
Grain per Lb. Gain	3.384	5.343	9.560	3.750	9.801	5.062
Hay per Lb. of Gain	1.748	8.370	4.168	1.813	4.191	2.415
Total Grain Eaten	2493.44	205.73	2643.48	1805.50	4899.86	4258.55
Total Hay Eaten	1288.12	322.53	1152.18	872.83	2095.58	2031.13

The Permanent Effects of Retarded Growth

Feed and Weight Record of Two Steers from 12 to 38 Months
Same animals as in table on page 269

Each animal was fed a full ration from the age of 12 to 38 months

	Age 12 to 38 Months		Age 32 to 38 Months		Whole Period 1 to 38 Mos.	
	No. 527	No. 529	No. 527	No. 529	No. 527	No. 529
Initial Weight (lbs.)	901.8	213.5	1687.4	1313.4	165.0	175.0
Final Weight (lbs.)	1825.0	2537.8	1825.0	1537.8	1825.0	1537.8
Gain during Period	923.2	1324.3	137.6	224.4	1,660.0	1362.8
Average Daily Gain	1.183	1.697	0.764	1.246	1.456	1.195
Grain Fed Daily	14.117	13.250	14.133	14.654	11.843	9.246
Hay Fed Daily	6.033	5.833	4.959	5.134	5.258	4.274
Grain per Lb. Gain	11.924	7.804	18.488	11.755	8.133	7.734
Hay per Lb. Gain	5.098	3.436	6.487	4.119	3.611	3.575
Total Grain Eaten	11,008.75	10,335.52	2543.98	2637.77	13,502.19	10,541.25
Total Hay Eaten	4706.68	4550.08	892.63	924.47	5994.80	4872.61

destroyed by the early stunting resulting from insufficient food. The gains, the amount of food consumed, and the gain in live weight based on each pound of grain consumed is shown in the tables for the various periods of the feeding experiment. (Tables on pages 269, 270.) It will be observed from the figures given in these tables that the animal subjected to severe conditions of feeding during its early life evidently had a greater capacity for growth, and made better use of the food consumed, than did the animal that received a generous ration during the same period of its life. There can be no question but that under certain conditions a significant check to the development of an animal may actually increase the rate of growth during the later periods of its life.

256. Climate. — Small animals like mice and domestic rabbits reared in artificially heated temperatures are noticeably larger than normal size and have less hair on the bodies. Cattle that are well fed and housed in warm barns have much less hair than animals of similar breeding that are required to live exposed to the rigors of severe cold weather.

257. The age factor in animal-breeding. — Very large numbers of farm animals are bred and produce offspring while still immature. The use of young sires among all classes of animals is general, but especially is their use common in the breeding of hogs, sheep and cattle. Various opinions exist among breeders as to the good or evil effects which follow this practice. Some of the evil results which it is claimed have followed the long-continued mating of immature animals are, a gradual decrease in the size of the breed; weakness of constitution; loss of prepotency in the transmission of valuable qualities;

a retardation of the growth of the young parents and, in some cases, a permanent dwarfing of the mother.

The supposed evil results following premature pregnancy must result either from variations produced in the fundamental constitution of the germ-plasm or in changes in the soma- or body-cells. Those breeders who contend that long-continued breeding of immature animals results in actually decreasing the size of the race or breed believe that the real character of the breed has been changed, and changed in such a way that the individuals of the breed are no longer able to produce offspring which possess the capacity to develop into animals of the recognized standard size of the breed. They would insist that the fundamental nature of the breed has been changed in such a way as permanently to affect their hereditary character. It is obvious that if this contention of breeders is correct, it is contrary to our present ideas of the inheritance of acquired characters. Biologists generally would first insist on a more accurate demonstration of the alleged fact that the size of the breed is actually diminished as a result of early pregnancy and lactation. If it should be found that this practice has apparently resulted in a smaller breed, the biologist would undertake to explain the observed fact on different grounds.

258. Premature breeding decreases size. — First, is it true that premature breeding has ever resulted in decreasing the size of the breed? It must be admitted that experienced breeders are very often accurate observers. Investigators who disregard the facts which have been determined by practical breeders through a long period of successful experience are neglecting a valuable source of knowledge on many of the complex problems of heredity and development. That breeders have observed that long-continued early breeding results in decreasing the size of the breed is indicated from an investigation made by J. M. Jones under the writer's direction in 1912. Specific questions were formulated and sent to a large number of successful breeders in America and Great Britain. From these replies it was determined that 216 breeders had observed that early mating resulted in weakening the breed, diminishing the fecundity, and decreasing the size. That such results followed was denied by thirty-five breeders. Of the total number replying, 158 believed that the size was decreased but that fecundity was not diminished. It is very clear from the statistics presented and from the extended replies of the intelligent breeders that under certain conditions the size of the animals comprising the herd of a breeder who continually breeds his animals prematurely is smaller than the size of animals in the herds of breeders who cause their animals to mate at a more mature age.

259. Decreased size due to early breeding not inherited. — In recognizing the fact that premature breeding does decrease the size of the breed under certain conditions, it is not necessary for us to assume that the tendency to decreased size in this case is inherited. Early breeding can have no direct influence in changing the fundamental constitution of the germ-plasm. It cannot, therefore, change the general prepotency of the breed in transmitting the recognized standard qualities of the race. We must therefore look for an explanation outside of the supposed influence on the hereditary powers of the breed. The real effects from premature mating are to be found in the development of the individuals as affected by environment. The effects of gestation and lactation, if observable at all, would be exhibited in the young parents or their offspring, or in permanent effects on the race or breed. It might, in fact, influence all three.

260. Influence of early pregnancy on the mother. — The chief harmful effect which follows the mating of immature breeding animals is a retardation or sudden check to the growth of the young mother. Investigation shows that the premature exercise of the breeding functions in the young female acts in some instances as a temporary inhibitor of growth. The author has for six years compared the growth of immature mothers with the normal growth curves of mothers mated when more mature. In this investigation six Duroc Jersey sows were divided into three groups. These groups of two sows each were designated respectively as immature, half mature and mature. The sows of the immature group were bred at the beginning of puberty or at the first heat. At this early period the young mother was very immature. The half mature sows were bred at about eighteen months and the mature sows at about thirty months of age. Careful records of the food consumed, the gain in live weight, and increase in body measurements were made of each animal in the experiment. Similar records were made of the female offspring of the original sows for several generations. The body measurements were taken with a view to determining muscular and skeletal increase. A large number of measurements was made, and these clearly demonstrated the fact that the early exercise of the breeding function in swine results in temporarily checking the growth of the mother. The investigations have progressed far enough at this time to measure also the ultimate effect upon the mature mother. While the observations have not yet included enough animals to justify us in speaking with finality, yet it seems entirely safe to conclude that under certain conditions premature breeding results in a permanent reduction in the normal size of the mother. Young sows which have been bred at the beginning of puberty and twice a year thereafter until full maturity have in every case been smaller at maturity than sows in the half mature and mature groups.

RETARDATION OF GROWTH DUE TO PREMATURE BREEDING OF SOWS

Measurements taken at 42 months of age

	LENGTH OF BODY	HEART GIRTH	HEIGHT AT WITHERS	DEPTH OF CHEST
	Centimeters	Centimeters	Centimeters	Centimeters
Immature [1] (Factor 6)	106	125	65	41
Half Mature [2] (Factor 3)	124	142	70	47
Mature [3] (Factor 8)	116	152	71	49

In the above table are recorded the more important measurements of a typical representative of each of the three groups. It will be observed that at forty-two months of age the immature sow was materially smaller in length of body, height at withers, depth of chest, and in heart girth. At this age (three-and-one-half years) the young sow had produced thirty-nine pigs in five litters, the half mature sow sixteen pigs in two litters, and the mature sow eight pigs in one litter. But one conclusion is possible

[1] Has produced 5 litters and a total of 39 pigs.
[2] Has produced 2 litters and a total of 16 pigs.
[3] Has produced 1 litter and a total of 8 pigs.

from these results, that early pregnancy and lactation does retard the growth of the young mother.

RETARDATION OF GROWTH DUE TO PREMATURE BREEDING

Measurements taken at maturity — 66 months of age

	LENGTH OF BODY	HEART GIRTH	WEIGHT AT WITHERS	DEPTH OF CHEST
	Centimeters	Centimeters	Centimeters	Centimeters
Immature[1] (Factor 6)	108	123	66	40
Half Mature[2] (Factor 3)	120	127	67	44
Mature[3] (Factor 8)	118	143	71	48

In the above table the measurements of the same animals at full maturity are shown. At the time these measurements were taken the parents were about five-and-one-half years of age. The sow that was mated at the beginning of puberty and bred regularly thereafter, producing sixty-nine pigs in nine litters, was smaller at maturity than either of the other groups which were mated at an older age. This result has been secured in a number of similar cases. In other words, the breeding of sows at a young age not only results in a temporary check to their development, but tends permanently to decrease the size. This permanent decrease is not very marked and practically may not be of great importance. It is clear from the

[1] Has produced 9 litters and a total of 69 pigs.
[2] Has produced 6 litters and a total of 49 pigs.
[3] Has produced 5 litters and a total of 32 pigs.

records that one important practical result of early mating is that by following this practice a much larger number of young are produced during the lifetime of the parent. In producing hogs commercially, this advantage might easily overcome any disadvantage arising from a reduction in the size of the mother.

261. Gestation and lactation in relation to growth. — The two most important phenomena associated with reproduction which might have a measurable influence on growth are gestation and lactation. The period of gestation in the mammalian domestic animals is the period during which the embryo is developing in the uterus, from the fertilization of the egg until the young animal is sufficiently matured to carry on an independent existence outside the body of the mother. During this period the unborn animal increases rapidly in size, and within its tissues the processes of cell division, absorption and assimilation proceed with exceptional energy. The nutrition for the growth of the fœtus during this period is supplied entirely by the pregnant mother. The fœtus itself has no means of nourishing its own tissues. It is wholly dependent upon the mother. To all intents and purposes the young animal in the uterus is an organic part of the body of the mother. The fœtus is an enormous parasite nourished by the mother through the circulation.

It is a popular opinion among many breeders that gestation is an exhaustive period for the mother, that during this period the mother must not only provide for her own physiological needs, but in addition must supply the materials needed for the rapid development of the fœtus. It has been generally believed that because of these facts the period of gestation is a severe strain on the pregnant mother. If the mother herself has completed her growth or has approached maturity in her development, the changes which take place in her organization and which may be due to her pregnant condition will obviously not affect her growth. On the other hand, if the mother is young and growing rapidly, gestation might act as a check to growth if the physiological processes of the mother are necessarily and chiefly directed toward the development of the fœtus. The physiological processes concerned in the nutrition of the fœtus are somewhat complex and the interrelations between the mother and her unborn young not completely determined. The information available on the subject does not specifically answer the question. We know that the absorption of fats from the intestine proceeds at a more rapid rate during pregnancy. Increased amounts of fat in the liver cells also are associated with pregnancy.

There is a tendency to increase in body weight during gestation. The maternal body increases in weight independently of the increase in size of the fœtus and fœtal membranes as shown by Gassner[1] and confirmed by others. This increase in weight is common to the mature pregnant mammalian mother and is not confined to the young parent only. It is possible, therefore, that this increase might be due to increased fat in the body and not to any increase in the skeleton. If this were found to be true, it would tend to confirm current opinion as to the retarding influence of gestation on the growth.

262. The Missouri experiments. — At the Missouri Experiment Station[2] careful measurements have been made of a large number of immature pregnant sows and

[1] Marshall, "Physiology of Reproduction," chap. XI.
[2] Mumford, Bulletins 131 and 141, Missouri Experiment Station.

of sows not pregnant for the purpose of answering, if possible, the question whether gestation is a period of such severe physiological strain on the young mother that it stops normal growth. A large number of animals have been under observation in this experiment and the results are fairly uniform. Measurements included records of changes in body weight, heart girth, length of body, height at withers, and other measurements.

The investigation is not complete, and further work may modify the conclusions which seem fully warranted at this time. The results so far justify the following conclusions:

1. The exercise of the reproductive functions, continuously from the beginning of puberty to full maturity, permanently decreases the normal adult size of the mother.

2. This permanent effect on the size of the mother occurs even under a favorable environment.

3. The period of gestation is not a check to growth when a full ration of nutritious food is supplied. The rate of growth is not lessened during gestation. There is some evidence that pregnancy is an actual stimulus to growth.

4. The period of lactation is a very severe physiological strain on the young mother, and during this period growth is apparently inhibited.

5. After the end of lactation or when the young are weaned, the rate of growth in the young mother is more rapid than before pregnancy and more rapid than in animals which have not been pregnant.

CHAPTER XIV

THE PRACTICE OF BREEDING

IF the modern breeds of domestic animals are compared with the original unimproved forms, remarkable differences will be observed. In many of the characters which have come to be of inestimable value to man, the modern animal is notably superior to the original unimproved forms. And in other characters, less substantial and economically significant, modifications of great scientific interest have resulted from systematic selection by man.

263. Improvement in size. — For many purposes the wild unimproved forms of the domestic animals are too small to accomplish successfully the work required by man. This demand for greater size and generally proportional increase in strength has led breeders consciously to select animals for size. The results of this selection may be observed in the gigantic modern draft horse. It is probable that all modern breeds of horses have developed from the same original type. The wild form was small in size, rough and somewhat angular in appearance, with short mane and scanty tail. From this animal we have through selection succeeded in producing the large, powerful draft horse with smooth, broad contours and heavy mane and tail. Contrasted with this type we have well-recognized pony types weighing less than 400 pounds. Both types undoubtedly descend from the same prehistoric form. Each type breeds true. The chief distinguishing character of size is firmly fixed in the germ-plasm and we must come to the conclusion that these radical differences have resulted from selection and have become firmly established hereditary characters. Similar differences in size among cattle, sheep and swine supply additional evidence that size is a character which may be radically changed through selection and that this variation may become so firmly fixed that it may be regarded as an established characteristic of the breed.

264. Improvement in function. — The most remarkable achievements in the improvement of the domestic animals are undoubtedly improvements in the various physiological functions of the animals useful to man. Some of the most noteworthy of these are the milking function in cattle, wool production in sheep, tendency to fatten in meat animals, speed in horses and egg-laying in the domestic fowls. A comparison of the productivity of each of these types of animals with the unimproved types gives ample evidence of the remarkable development which has taken place through the agency of man's selection.

265. The milking function. — The ability of mammalian animals to produce milk is closely correlated with the reproductive functions. The mammary glands function primarily for the purpose of supplying a nutritious and easily digested food for very young animals. Among wild forms this function persists only for a comparatively brief period and its continuance is determined by the needs of the young mammal. Under domestication the milking function in the domestic cow represents a remarkable improvement. The wild cow probably supplied milk to her offspring only four or five months. The amount of milk supplied by the wild cow is limited to a few pounds daily.

The milking function in the domestic cow has become through man's selection an hereditary character of the greatest importance. While in the domestic cow the milking function is still closely correlated with the reproductive functions, it is nevertheless developed to such an extent that it bears no very important relation to the needs of the young offspring. Certain individual cows have this function developed to such an extraordinary degree that they produce milk continuously without interruption for many years and often regardless of whether the cow becomes pregnant or not. (Plate XXVI.)

The Holstein Friesian cow, Duchess Skylark Ormsby (Plate XXVII), produced 27,761 pounds of milk in one year, containing 1205 pounds of fat. The Jersey cow, Sophie 19th of Hood Farm (Plate XXVIII), has a record of 17,557 pounds of milk containing 999 pounds of fat. The Guernsey cow, Murne Cowan, is officially reported as having produced 24,008 pounds of milk and 1098 pounds of butter-fat in one year. Garclaugh May Mischief, an Ayrshire cow, has a similar yearly record of 25,329 pounds of milk and 894 pounds of fat.

266. Improvement in wool production. — The first importations of Spanish Merino sheep into the United States were made about 1815. At that time the average weight of fleece was three or four pounds. The weight of fleeces of American Merinos increased gradually from that time until 1885. Authentic records of single fleeces weighing from thirty to forty pounds are now available. The Oklahoma Agricultural College has reported that the two-year-old Rambouillet ram Loraine owned by that institution has produced a fleece weighing 46 pounds which is claimed to be the heaviest fleece ever taken from a single sheep. The staple of this fleece was three and one-fourth inches in length, which when straightened measured five inches. It is interesting to note in this case that the fiber was of unusual fineness, averaging $\frac{1}{1800}$ of an inch in diameter. According to Hunt,[1] the average weight of fleece of all sheep in the United States in 1850 was 2.4 pounds a head. In 1900 the average weight of fleece was 6.9 pounds a head. This remarkable development in the improvement of the wool-producing qualities of animals must be wholly credited to the skill and enterprise of the American shepherd.

267. Improvement in tendency to lay on fat. — It is more difficult to give accurate statistics showing the improvement in meat-producing animals. At the beginning of the eighteenth century, fat animals were not placed on the market until they were four or five years of age. Even as late as 1875, three- and four-year-old fat animals were the most common types of cattle to be found in the fat stock markets of America. By careful selection the tendency to lay on fat has been developed in animals to such an extent that now it is common to find young animals carrying fat equal to that shown by four- and five-year-old bullocks of earlier years. This tendency to lay on fat at an early age is transmitted through heredity. The amount of food required to make one pound of gain is much less in the younger animals and the improvement in this respect, therefore, is of great economic significance. It requires less food to-day to produce the fat beef, pork and mutton sold in the markets of the World than was the case before the improvement of this tendency to early maturity. (Plates XXXI and XXXII.)

[1] Hunt, " Cyclopedia of American Agriculture," vol. 3.

268. Improvement in speed. — The ability to go fast at the trot is a development of American horse-breeding enterprise. The first trotting race was held at Boston in 1815. The fastest time made in this race was a mile in three minutes. As the result of interest in trotting races and the invention of light horse-drawn vehicles, the demand for speed at the trot resulted in great improvement in this direction. The record time of trotting horses decreased from year to year until at the present day (1916) several horses have been developed that are able to trot a mile in less than two minutes. The examples mentioned are all noteworthy as examples of man's power to change fundamentally the form and function of the domestic animal. These examples could be indefinitely multiplied.

269. Selection. — All wild forms that have been domesticated possess characters of economic value to man. They were domesticated for that reason. These valuable qualities have been, in many cases, greatly improved through the agency of man. So great has been the improvement in many animal forms that the domesticated animal is markedly different from his wild ancestors. Many varieties, races or breeds exist and each of these differs in important particulars not only from its wild relatives but from all other varieties having similar ancestral history.

How have these valuable characteristics come into existence? What natural laws have guided man in the improvement of animals? Are the valuable characteristics of animals due chiefly to inheritance or are they in most part the result of improved conditions which surround the domestic animals? Is it possible for us from a study of the history of the achievements of animal-breeders to establish a guide to future practice in the improvement of animals? Answers to these questions not only have great value to practical breeders but have profound biological significance.

270. Natural selection. — The theory of natural selection which has so long influenced and determined the trend of zoölogical thought is an attempt to explain how the qualities of animals in nature have come to be. Through natural selection organic beings tend to adapt themselves to their surroundings. The more nearly the qualities of animals are favorable to a successful existence under the conditions surrounding the individual, the more certainly will the animal live and reproduce.

The insect that rests upon the branches of trees will escape insectivorous birds more certainly if the color of the body imitates that of the bark upon which it rests. The color of the jungle animal blends so completely into the general landscape that its presence is not detected by its enemies.

The shore birds or waders in their pursuit of food in the shallow waters of the shore have developed long legs. The giraffe, feeding as it does upon the leaves on the branches of trees, has found a long neck to be a desirable quality in securing food. And in this case the longer the neck the more certainly will the animal survive when food is scarce.

Darwin assumed that under conditions similar to those mentioned, animals that vary slightly in the desired direction would be preserved and would reproduce, while those animals that varied away from the valuable quality would perish. An animal might continue to develop indefinitely in a given direction through continuous variation. Darwin also recognized the existence of discontinuous variation.

The origin of the peculiar characters which are valuable in the domesticated animal is believed to have come about through a similar process of selection but with this difference: Man has introduced methodical selection by which all animals are preserved, not only those suited to their particular environment but also and chiefly those which possess characteristics that are valuable to man. Many complex and difficult questions have arisen in connection with the improvement of the domesticated animals involving the most difficult problems of inheritance.

271. Methodical selection. — The selection practiced by man is known as methodical selection. Animals are selected because of their peculiar fitness for special purposes. The powerful, heavy draft-horse is the result of generations of careful selection and breeding in an effort to produce an animal that can pull heavy loads.

In the same way but with a different standard of selection, the speed horse has developed an extraordinary ability to run fast under the saddle or trot at a rapid pace before the carriage.

Similarly, the beef animal, the dairy cow, the mutton and wool sheep, the lard and bacon hog, the swift greyhound, the massive St. Bernard, the intelligent Scotch collie and many other types useful to man have come from methodical selection.

272. Importance of selection in animal-breeding. — Careful study of the methods which have been practiced by breeders of the domestic animals cannot fail to lead to the inevitable conclusion that selection has been the most important if not the chief principle followed in bringing about the present highly developed forms among animals.

Certainly not all the phenomena which are exhibited

in the practices and results of animal-breeders can be explained upon the basis of selection alone, but it is quite certain that without selection little practical use could be made of the known laws which govern the transmission of characters. Man's chief agency in the improvement of animals has not been a conscious effort to bring about a series of variations or mutations of a certain kind, but it has rather been in the direction of preserving such valuable characteristics as have been already in existence or have appeared through variation. These characteristics have been intensified by judicious matings and perpetuated as a result of the keen insight of the skillful breeder. The successful breeder has ever in mind an ideal. He is at all times alert to detect variations which approach this ideal.

Darwin believed that most of the improvement wrought in domestic animals was due to minute or continuous variations from the less desirable to the more desirable. This belief has been general among animal-breeders themselves. Working upon this assumption, the work of the breeder consisted merely in accurately observing the variations which tended in the desired direction.

But continuous variation assumes a perfect series of infinitesimal steps, each grading into the one higher or lower. Such continuity exists in the growth of animals from birth to maximum development. Continuous variation is also illustrated in the physical world by changes in temperature. In the history of the improvement of domestic animals there are many examples of marked and sudden variations in the offspring which cannot be continuous. Such variations have been called discontinuous variations or mutations. It is certain that much of the improvement in the domestic animals has come from valuable mutations which have been recognized by the alert breeder.

But whether the present valuable characteristics of the domestic animals have come through continuous variations or by sudden mutations or in any other way, selection by man has been the one outstanding fact in the development of the many valuable races and breeds of animals. It is conceivable and very probable that similar variations have been induced by similar causes in all wild forms, but wild forms have remained relatively stationery while the domestic races have been greatly improved.

273. Aids to selection. — The breeder has consciously or unconsciously brought to his aid numerous practices which have greatly facilitated his efforts. The manufacturer who has invented a labor-saving device or a complicated machine to perform certain work puts it to the test by actually applying the machine to the work in hand. Likewise the skillful breeder has found it greatly to his advantage to test his animal creations by actual performances.

The breeders of trotting horses have no means of determining how successful they are in producing speed except by trial on the track and it is not only necessary to train now and then a horse but every individual in the stud must be tested to be certain that all possess the quality of speed.

The highly successful breeders of meat animals, cattle, sheep and swine, maintain their breeding animals in a high state of condition. In some cases the meat animals used for breeding purposes by noted breeders are kept so fat as to interfere with the normal reproductive functions. But even so, such a practice is essential to the skillful breeder because in no other way can he determine whether or not his breeding animals themselves possess the proper tendency to lay on fat which is an essential characteristic of a well-improved meat animal.

A high condition of the breeding animals does not in any sense give the beef animal a greater power to transmit the tendency to lay on fat to the offspring, as some breeders have believed, but it does give him a selective device or measure by which he may always know whether his breeding animals themselves possess the characters which it is desired to transmit to the offspring.

The breeders of dairy cattle feed their cows to full capacity and surround them with every favorable condition for the maximum production of milk. This practice gives the breeder the only accurate measure of individual performance and places in his hand a selective device by means of which he can quickly eliminate from his herd those animals that have not inherited the capacity for high production. There is no other accurate measurement which the breeder can employ that will certainly register his progress in the improvement of his favorite breed.

274. The real results of selection in the improvement of the domestic animals. — The many improved breeds of live-stock which are now possessed of qualities of great economic value to man, have come to this possession through artificial selection practiced by man. Are these qualities which distinguish the improved breed from the wild form a permanent possession of the race? Have the selective processes applied by man resulted in so fixing the desirable heritable characteristics that the breeder may depend upon inheritance to repeat the characters of the parents in the offspring? There are some individuals in each breed that represent near perfection in the development of characters. There are many others that are more or less deficient in such characters while other individuals are actually mediocre. Will the individual with near perfect characters reproduce other individuals of the same high quality? Will the mediocre sort beget like mediocrity? In other words, are the qualities which have resulted from artificial selection indelibly impressed on the hereditary substance of the germ-plasm in such a way as to be certainly transmitted to the offspring? Exact answers to these questions cannot be given without full knowledge of all the conditions, but recent investigations have thrown much light on these important questions.

275. Selection within pure lines. — Selection as practiced by man has undoubtedly resulted in marked improvement of the domestic animals. This improvement is apparently transmitted, or at least it is possible by continued selection to reproduce desirable characteristics. By continually selecting individual breeding animals which show a tendency to vary in the direction desired, it has been assumed that the race or breed would gradually move in the direction of selection.

At one period in the history of Shorthorn cattle, all breeders were agreed that their improvement was mainly to be accomplished in the direction of increased size. Consequently the animals of greatest scale were selected and mated with the apparent result that the average size of the breed was increased. But actually what was accomplished after many generations of selection? Was a new breed created? Had selection acted as a causative principle? The teachings of Darwin and his followers has undoubtedly resulted in creating the impression among many students that in this case a new breed had actually been created with new characters resulting from selection of continuous variations. That the results obtained in this and similar examples may be explained on other and more probable grounds has been clearly demonstrated by Johannsen.[1] This Danish botanist working with garden beans found that the whole population is made up of many races which he called "pure lines." Selecting from populations composed of pure lines, the breeder merely sorts out one of these races and in time secures a pure line. This pure line or race is not new. It has not resulted from gradual or continuous variations but has simply been separated out from a number of others. According to Johannsen, selection within a pure line during a number of generations had no effect in improving the variety. The germplasm of all individuals of a pure line tends to become homogeneous.

The mating of individuals from a pure line having in their respective germ-plasms identical factors will result in producing identical offspring and hence any amount of selection will prove futile. Such pure line individuals are much more likely to be found among plants than among animals. Plants that are self-fertilizing may be expected to develop typical pure lines and selection within such pure lines will be of no avail. Among the higher animals similar results undoubtedly occur, but the difficulties of securing a strictly pure line are very much greater owing to the necessity of mating two distinct individuals whose ancestry, while similar, cannot possibly from the very nature of the case, be exactly identical.

[1] Johannsen, "Elemente der exakten Erblichkeitslehre," 1909.

276. Vilmorin's pure line wheat-breeding. — A good example of the pure line theory among plants is to be found in the very practical and meritorious work of Louis de Vilmorin, who began the improvement of commercial varieties of wheat in France about 1840. Vilmorin carefully selected a single head of outstanding merit and from this by in-breeding established a pure line or variety which has bred true to the original ear or head. The commercial seed offered by Vilmorin was always descended from a single plant. Selection in this case failed to bring about any improvement over the original plant, as shown by the Hagadoorns.[1]

"In 1911 Mr. Meunissier, the genetist of the firm of Vilmorin, found the collection of original ears of the varieties with which Louis de Vilmorin half a century back began his living museum. Some of these wheats are from the harvest of 1843, others date from 1850, or intermediate dates. Mr. Meunissier chose three dozen ears of varieties which are still in the collection, and which have therefore been bred continuously as pure lines for about fifty years. We compared these ears to ears of the 1911 harvest, and photographed them side by side. Some of these pairs of ears are here shown, each pair consisting of the old ear, and its descendant, half a century later. All these generations of selection have not changed any one of the varieties one little bit. It can therefore safely be concluded from this series of experiments, that selection can have no effect in material pure for its genetic factors. Genetic factors are constant."

277. Selection most useful when genetic factors are not pure. — The pure line theory explains why any

---

[1] Mrs. C. and Dr. A. L. Hagadoorn, "Selection in Pure Lines," *American Breeder's Magazine*, 1913, p. 165.

improvement among the best strains of animals is so difficult. Highly improved breeds or families among the domestic animals have reached a degree of purity wherein the germ-plasm represents an approach to homogeneity. When such a pure line has been established, the old adage that "like begets like" becomes a really working principle and a guide to practice. The novice need not expect to accomplish great improvement in the highly developed breeds of live-stock. The greatest improvement will be made in breeds or strains of mixed breeding, genetically speaking.

The improvement of animals of mixed character which will result from the efforts of a breeder is probably to be explained as a gradual separation of the desirable pure lines and mating the individuals which possess these characters. This results in time in the establishment of a strain or breed which is prepotent in transmitting the desired qualities when after many generations of pure line breeding this strain has reached a point where the germ-plasm of the breeding animals is pure in respect to its genetic factors; then any considerable improvement will be no longer possible.

278. Pure line theory not opposed to improvement by selection. — The practical breeder whose experience has demonstrated clearly that improvement among the registered or "pure-bred" races of the domestic animals has followed careful and persistent selection will hesitate to accept the statement of Johannsen and his followers that "selection within the pure line is without effect." But the improved breeds of live-stock, however pure in breeding, are not to be regarded as fulfilling the requirements of a "pure line" in the biological sense. The breeds of live-stock are yet far from homogeneous in the sense that they have been selected to a point where the germ-plasm of different individuals of the breed, is identical in composition. The pure line of the biologist is after all a purely imaginary conception. It is conceivable that such a condition may be produced among self-fertilizing plants, but among the higher animals it is certainly true that no such germinal purity has yet been attained. The practical animal-breeder may still continue to hold fast to selection as his chief means of improvement with the assurance that in all the higher domestic animals we have not yet reached the ultimate limits of improvement. The animal-breeder has not yet produced a pure line, biologically speaking, and has not, therefore, reached a point in his breeding when it can be accurately said that selection is without effect.[1]

279. Pedigree. — The pedigree of an animal is a record of the ancestors. It is a valuable historical document. If it includes the names of many animals of outstanding quality, it is a good pedigree. If the ancestors were mediocre individuals of no special merit, the pedigree is inferior. The mere fact that an animal is recorded in a recognized herd-book does not signify that such animal has a good pedigree. The careful breeder must have a thorough knowledge of the history of the breed and especially of the ancestors of the individual animal under examination.

The immediate ancestors are most important. A noted sire or dam appearing in the pedigree six or eight generations back is of far less importance than one which appears in the first, second or third generations. Many breeding animals are sold on the strength of the fact

[1] Castle, "Pure Lines and Selection," *Journal of Heredity*, 1914, p. 93.

that they are descended from some famous sire or dam ten or even twenty generations back. If Galton's law of ancestral heredity is a fair estimate of the potential strength of a breeding animal, we should expect that the individual animal would inherit one-half from his two immediate parents, one-fourth from his four grandparents, and one-eighth from his eight great-grandparents. From one great-grandparent, therefore, we should expect the inheritance in the descendant to be represented by the fraction $\frac{1}{64}$. It is evident that a single ancestor six or eight generations removed would contribute a very small fraction of the sum total of qualities of the individual.

In the application of Galton's principle, it is necessary to assume that all the ancestors are equally prepotent. This assumption is contrary to the experience of breeders. Certain individuals are known to be more prepotent in certain characters, or, more properly speaking, certain characters are dominant and others recessive. The dominant characters will largely determine the character of the offspring. It is conceivable that a dominant character which it is greatly desired to perpetuate in a strain might reappear in the offspring through many generations. It might be argued that because of this fact a study of the characters of an ancestor far removed having this character would be valuable. To this it must be said that if the character is dominant it will be clearly apparent in the immediate ancestors. If it is not apparent it may have been lost or overshadowed by the development of other characters, and if so it is good evidence that the real character of the strain is being determined by nearer and stronger ancestors. Every modern biological conception gives added weight to the principle that it is the recent ancestors that should be most carefully investigated by the practical breeder if he is to obtain any valuable basis for estimating the breeding powers of an individual.

280. Registered breeding animals. — In purchasing a breeding animal, the breeder desires some guarantee of purity of breeding. Every recognized breed, therefore, maintains an association of breeders organized for the purpose of safeguarding the purity of the breeding animals and advancing the general interests of the breed. Each association maintains a record book in which every animal recognized as a pure-bred animal of the breed is recorded. Each association has its own rules governing the registration of animals. In practically all breed associations, the offspring of registered parents can be recorded. There are few exceptions to this rule. In some associations animals may be registered upon the basis of their performance. The Standard Bred or American Trotting Horse Registry Association will record any animal that has made an authentic record of a mile in two minutes and thirty seconds on an approved track and under regulation rules. A few associations in the past have permitted the registration of animals having a certain number of top crosses to registered sires. The history of all breed associations is similar. A few animals of similar characters have attracted attention because of their peculiar value for certain purposes. These animals have been interbred and gradually a family or strain developed which excels in certain valuable characteristics. The admirers of this family or strain have organized to preserve the strain. Experience has shown that one of the first and most useful steps is to record the individual breeding animals of outstanding merit. The descendants of these animals constitute the breed, and ultimately only descendants of animals recorded in the registry book are eligible to registration.

281. Registry associations. — The recognized registry associations of the United States at present (1916) are listed below: [1]

AMERICAN HORSE RECORD ASSOCIATIONS

Arabian Horse Club of America.
American Association of Importers and Breeders of Belgian Draft Horses.
Cleveland Bay Society of America.
American Clydesdale Association.
French Coach Horse Society of America.
National French Draft Horse Association of America.
German, Hanoverian and Oldenburg Coach Horse Association of America.
American Hackney Horse Society.
American Morgan Register Association.
Percheron Society of America.
The American Breeders' and Importers' Percheron Registry Company.
American Saddle Horse Breeders' Association.
American Shetland Pony Club.
American Shire Horse Association.
American Suffolk Horse Association.
American Trotting Register Association.
The Jockey Club.
The Welsh Pony and Cob Society of America.

AMERICAN JACKS AND JENNET RECORD ASSOCIATIONS

American Breeders' Association of Jacks and Jennets.
Standard Jack and Jennet Registry of America.

AMERICAN CATTLE RECORD ASSOCIATIONS

American Aberdeen-Angus Breeders' Association.
Ayrshire Breeders' Association.
Brown Swiss Cattle Breeders' Association.

[1] Data from United States Department of Agriculture.

American Devon Cattle Club.
Dutch Belted Cattle Association of America.
American Galloway Breeders' Association.
American Guernsey Cattle Club.
American Hereford Cattle Breeders' Association.
Holstein-Friesian Association of America.
American Jersey Cattle Club.
American Kerry and Dexter Cattle Club.
Polled Durham Breeders' Association.
American Polled Hereford Breeders' Association.
Red Polled Cattle Club of America, Inc.
American Shorthorn Breeders' Association.
American Dairy Shorthorn Cattle Club.

AMERICAN SHEEP RECORD ASSOCIATIONS

American Cheviot Sheep Society.
American Cotswold Registry Association.
The Continental Dorset Club.
American Hampshire Sheep Association.
American Leicester Breeders' Association.
National Lincoln Sheep Breeders' Association.
American and Delaine Merino Record Association.
Dickinson Merino Sheep Record Company.
National Delaine Merino Sheep Breeders' Association of Washington County.
Standard Delaine Merino Sheep Breeders' Association.
American Rambouillet Sheep Breeders' Association.
International Von Homeyer Rambouillet Club.
Michigan Merino Sheep Breeders' Association.
Vermont, New York, and Ohio Merino Sheep Breeders' Association.
American Oxford Down Record Association.
American Romney Marsh Breeders' Association.
American Shropshire Registry Association.
American Southdown Breeders' Association.
American Tunis Sheep Breeders' Association.

GOATS

American Angora Goat Breeders' Association.
American Milch Goat Record Association.

AMERICAN SWINE RECORD ASSOCIATIONS

American Berkshire Association.
American Large Black Pig Society.
Cheshire Swine Breeders' Association.
O. I. C. Swine Breeders' Association.
Chester White Record Association.
American Duroc Jersey Swine Breeders' Association.
National Duroc Jersey Record Association.
American Hampshire Swine Record Association.
American Poland China Record Company.
National Poland China Record Company.
Standard Poland China Record Association.
American Tamworth Swine Record Association.
American Yorkshire Club.
National Mule-foot Hog Association.
Mule-foot Hog Breeders' Association.
American Mule-foot Hog Record Company.

282. Community breeding. — The establishment and maintenance of a recognized breed is a coöperative enterprise. No single individual can alone successfully establish and maintain a breed. Other things being equal, the larger the number of breeders engaged in the improvement of a given breed, the more certainly will the breed be improved and established on a permanent foundation. It is also true that it is distinctly to the advantage of a breed to be owned by breeders living in the same region. The interests of the breeders are greatly enhanced if many workers in a restricted area are breeding the same class of animals. This is particularly true in the case of the person who maintains a small herd or flock. The economic value of establishing for a community the reputation of breeding a very large number of a certain breed has been demonstrated in many localities in this and other countries. Recognizing these advantages, educational organizations and breed associations have encouraged community breeding enterprises. In some localities this has taken the form of coöperation among three or four small breeders in the purchase of a valuable bull. In the United States many associations have been formed among farmers for the purchase of a valuable stallion. Even in the absence of conscious coöperation for improvement, if a large number of the farmers in a given neighborhood are like-minded in the selection of breeds and all produce the same breeds, each particular animal will actually be more valuable because prospective buyers will be attracted by the opportunity for selection where large numbers are available.

283. Importance of numbers. — The present high quality of the highly improved and valuable breeds of the domestic animals has been the result of long-continued and rigid selection. The perpetuation of the improved characters already obtained rests upon the opportunity for continued selection of the same kind. The effectiveness of selection will depend upon the number of individual animals which are concerned in any given breeding project. It follows, therefore, that the breeder who produces large numbers has a decided advantage over the one whose opportunity for selection has been confined to a relatively small number of animals. A noted breeder of dogs who was asked to give the secret of his success replied, "I breed many and hang many." The breeder whose operations are limited to a relatively small flock or herd cannot expect to accomplish as much in the improvement of any class of animals as the breeder handling much larger numbers.

284. Selecting the best. — The improvement of animals has come chiefly through selection. In the actual process of selection, men have followed various methods with the ultimate purpose of obtaining finally a race or breed of fixed characters, that is, characters which are represented by definite determiners in the germ-plasm in such a way that the individual animals of the breed are able to transmit these desirable characters to their offspring with a reasonable degree of certainty. Thus many breeders have surrounded their animals or plants with exceptionally favorable conditions and have selected those which have developed the prized qualities most perfectly under such conditions. This was the method of Hallet in developing improved varieties of wheat. Undoubtedly many of the early breeders based this practice upon a very deep-seated but mistaken belief in the inheritance of acquired characters. This method has been very successful in a considerable number of cases, both among plants and animals, but not because the environmental factors involved had fundamentally changed the real character of the germ substance. In all such cases the improved environment acted merely as an efficient selective device and indicated those individuals which actually possessed the capacities valued by the breeder. This method has often failed in accomplishing lasting improvement, because the conditions surrounding the breeding stock are not average conditions and the apparent improvement may be wholly due to a better food supply or more room and not due to fundamental differences in the germ. Another method of improvement which involves selection of the best is to place the plant or animal under ordinary or even unfavorable conditions and select those individuals which appear best able to develop the desired qualities under such conditions.

285. Selecting chance variations. — We know that sudden and important variations often occur in the germ-plasm. Some of these variations may be of such a character as to have great economic value to man. Variations of this character are invariably transmitted, and the wise and observant breeder may often make rapid progress by making such chance variation the basis of his selections. This is the method of De Vries. In following this method the breeder does not consciously undertake to cause variation but rather to take advantage of those which have resulted from natural causes. The breeder of the domestic animals will often have difficulty in determining whether a given variation is due to environmental causes or is due to fundamental changes in the germ-plasm. The skillful breeder, however, will conclude with reasonable assurance that when an individual animal exhibits a rare and unusual aptitude in the development of a certain character or characters, in a herd in which the individuals are all maintained under identical conditions, the rare development may be regarded as a germinal variation. Such germinal variations may under certain conditions become the foundation of a new strain.

286. The Burbank method. — If the selection of variations is the road to success in improvement, then why not systematically attempt to cause variations and thus increase the chances for discovering á desirable mutation? This is the plan followed by Burbank. By crossing a great number of individuals, variations are secured, and by a process of gradual elimination the outstanding variants are retained and reproduced. This plan does not create any new forms, but depends on the well-known tendency of unit characters to rearrange themselves in new combinations which for all practical purposes may really become a new creation. This method involves the propagation of the improved individual by budding, grafting or similar asexual method and cannot therefore be successfully applied in animal-breeding. This method, like all the others described, is open to the objection that it is after all based on mere chance. It is empirical and unscientific. The proportion of failures to successes is too great, and for these reasons it is a slow process of improvement.

287. The mendelian method. — The mendelian method is based on the law of dominance and the segregation of unit characters. The first step is to determine by the behavior of the desired character in transmission whether it is a dominant or a recessive character. If it is recessive, then it is only necessary to combine two recessives, as recessives are homozygous and always breed true. If the desired character proves to be a dominant, it is first necessary to determine whether it is present in a heterozygous or a homozygous condition. If it is homozygous, it will breed true. If it is heterozygous, by in-breeding and gradually eliminating the recessives it is possible greatly to increase the number of dominants appearing and thus practically establish a pure strain. This method is also more successful in plant- than animal-breeding. The animal characters which have come to be recognized as of value to man are generally complex and do not behave in transmission as unit characters.

INDEX

Abortion, 120.
    a cause of sterility, 120.
    agglutination test for, 128.
    contagious, 123.
    diagnosis of, 127.
    treatment of, 124.
Acquired characters, 157–160, 162.
    examples of, 162.
    influenced by food supply, 162.
Acquired diseases, 180.
Age and fecundity, 96.
    of poultry affects fertility, 96.
    of ram influences fertility, 94.
    of sheep influences fertility, 93.
    of swine influences fertility, 91.
Age factor in animal-breeding, 271.
Agglutination test for contagious abortion, 128.
Anaphase in cell division, 12.
Arkell, 187.
Artificial insemination methods, 44.
Asexual reproduction, 16.
Ash in ration, effect on fœtus, 263.
Atwood, 94.

Bachhuber, 50.
Barrenness, *see also* sterility, 119.
Bateson, 210.
Beard, 76.
Beinn Bhreagh flock, fertility of, 99.
Bell, 99.
Berberrich, 89.
Birth, number of young, 85.
Birthweight of lambs, 262.
Bitch, genital organs of, 32.
Blending inheritance, 135.
Blue-gray cattle, 247.
Border Leicester sheep, 87.
Boyd, 249.
Breeding prematurely decreases size, 272.

Breeding season, 52.
    diœstrum, 55.
    metœstrum, 55.
    œstrum, 55.
    proœstrum, 54.
Brooks, 134.
Brown-Sequard experiments, 208.
Brull, 127.
Burbank method, 302.

Calcium in rations for pregnant swine, 264.
Castle, 194–294.
Castration, 22.
    influences secondary sexual characters, 186.
Cattalo, 250.
Cell, 1.
    contents, 4.
    division, 8, 9, 10, 11.
    germ, 2.
    growth, 7.
    structure, 4.
    theory, the, 1.
    the physiological unit, 3.
Cell division a cause of variation, 210.
Characters correlated with fertility, 102.
    originate in germ-plasm, 195.
Chillingham cattle, 224.
Chromatin, 6.
Chromosomes, 12, 13, 135.
Cole, 50.
Color-blindness, 188.
Community breeding, 299.
Connaway, 128.
Contagious abortion, 123.
    complement fixation test, 128.
    diagnosis of, 127.
    treatment of, 124.

Continuous and discontinuous variations, 151.
Controlling sex, 189.
Cornevin, 229.
Corn ration, effect on fœtus, 266.
Cow, reproductive organs of, 29.
Cows, exceptional fertility of, 107.
Cross, the first, an improvement, 247.
Cross-breeding, 243.
　a cause of variation, 248.
　advantages, 244.
　effect on breeding powers, 244.
　influences fertility, 105.
　permanent results from, 243.
　to increase fertility, 245.
　to increase size, 246.
　to restore constitution, 246.
Crossing and heredity, 247.
　bison and cattle, 249.
　species, 249.

Dalrymple, 125.
Darbyshire, 149.
Darwin, 52, 72, 90, 146, 198, 210, 221, 223, 225, 285.
Davenport, E., 100, 210.
Decreased size from early breeding not transmitted, 273.
Development, 255.
De Vries, 152, 302.
Diagnosis of contagious abortion, 127.
Di-hybrids, 155.
Disease, 179.
　acquired, 180.
　congenital, 181.
　predisposition, 181.
Duration of lactation, 81.
　of œstrum, 63.

Early maturity, 283.
　pregnancy, influence on mother, 274.
Eckles, 122.
Evvard, 263.
Ewart, 122, 168.
Exercise, effect on lactation, 83.
　favors fertility, 111.
Experiments by Mendel, 131.

Factors affecting fertility, 100.
Fallopian tubes, 27.
　obstruction of, 118.
Fatness, excessive, unfavorable to fertility, 99.
Female reproductive organs, 24.
Fertility, 85.
　characters correlated with, 102.
　confinement unfavorable, 89.
　correlated with size, 86.
　domestication increases, 87.
　duration of reproductive period influences, 88.
　exceptional in cattle, 107.
　　in horses, 106.
　　in poultry, 112.
　　in sheep, 110.
　frequency of recurrence of œstrum, 88.
　number of mammæ as related to, 101.
　nutrition, effect of, on, 98.
　Poland China breed, 92.
　relation of age to, 91–93.
　relation of gestation to, 87–88.
Fertilization, 33.
　changes in ovum resulting from, 36.
　nature of, 34.
First cross an improvement, 247.
Flushing ewe, 99.
Fœtal development and heredity, 261.
Fœtus, size and vigor influenced by ration, 266.
Food supply, excessive, causes sterility, 99.
　and body changes, 165.
　influence of restricted, 165.
Free-martin, 129.
Function, improvement in, 281.
　milking, 281.

Galton, 102, 295.
Gametic purity, 141.
Gentry, 234.
Germ-cells, 2, 12.
　origin of, 40.
Germ-plasm, 161.
　origin of new characters in, 206.

Gestation, 66.
    causes of variation in period, 72.
    period of, 69.
Geyelin, 97.
Goodale, 23, 187.
Goodnight, 249.
Graafian follicles, 25.
Grading, 244.
Growth, 256.
    a cell function, 259.
    by cell division, 7.
    capacity for, 257.
    development of fœtus, 260.
    effect of climate, 271.
    factors influencing, 257.
    impulse strongest in youth, 256, 260.
    influence of age, 271.
    influence of food supply, 257.
    influenced by early pregnancy, 276.
    relation of lactation to, 277.
    retarded, permanency of, 268.

Hagadoorns, 292.
Hallett, 147, 301.
Hart, 64, 266.
Heape, 42, 54.
Heat or œstrum, 41.
    duration, 63.
    during pregnancy, 60.
    effect of ration on, 64.
    recurrence, 63.
Helme, 75.
Heredity, 131.
    and development, 132.
    and fœtal development, 261.
    and sex, 183.
    and variation not antagonistic, 124.
    definitions, 132.
    kinds of, 134.
Hinny hybrid, 252.
Huish, 43.
Humphrey, 64, 266.
Huxley, 151.
Hybrids, 249.
    cattle-bison, 249.
    cattle-zebu, 253.
    hinny, 252.

Hybrids — *Continued*
    mule, 170–250.
    sheep-goat, 254.
    zebra-horse, 253.

Immunity, 181.
Improvement, 280.
    in early maturity, 283.
    in function, 281.
    in milking function, 281.
    in size, 280.
    in speed, 284.
    in tendency to lay on fat, 283.
    in wool production, 282.
In-breeding, 217.
    advantages claimed, 218.
    bad results from, 219.
    Berkshires, 234.
    cattle, 223.
    Darwin's researches, 231.
    decreased fertility following, 220.
    definition, 217.
    dogs, 228.
    fixing characters by, 240.
    limits of, 238.
    loss of vigor from, 220.
    mice, 231.
    pigs, 226, 234.
    prepotency of in-bred animals, 241.
    researches of Ritzema Bos, 232.
    results with different species, 242.
    selection important, 238.
    sheep, 227.
    Wistar institute experiments, 233.
Incubation, fowls, 74.
Inheritance, 131.
    alternative, 135.
    blending, 135.
    mendelian, 137.
    mosaic, 136.
    of disease, 179.
    of polled character, 144.
Insemination, artificial, 42.
Intoxication of male parent, effect on offspring, 49.
Iwanoff, 43.

Kehrer, 75.
Kempster, 97.
Kulbs, 89.

Lactation, 80.
　duration, 81.
　effect of exercise on, 83.
　habit, 82.
　heredity, 82.
　influence of climate, 83.
　mare mule that secretes milk, 83.
　œstrum, 59.
　relation of food supply to, 82.
　unusual lactation, 83.
Lambs, birthweight of, 262.
Langer, 31.
Law, 68, 117, 124.
　of ancestral heredity, 295.
　of dominance, 139.
　of segregation, 139.
Lead poisoning, effect on male germ-cells, 50.
Lillie, 47.
Lock, 133.
Lord Morton mare, 167.
Lovejoy, 59, 236.

McCollum, 64, 266.
MacFadyean, 124.
Male reproductive organs, 19.
Males, milk secretion by, 28.
Mammæ, number as related to fertility, 101.
Mammary glands, 28, 81.
　milk secretion by males, 28.
　milk secretion in, 31.
　structure of, 30.
Mare, genital organs, 26.
Marshall, 47, 53, 56, 96.
Maturation of ovum, 13.
Mendel, 136, 137, 138.
Mendelian method, 303.
Metaphase in cell division, 10.
Miles, 200.
Milk, 80.
　analysis of milk from mare mule, 82.
　secretion by males, 28.
　secretion in mammary glands, 31.
Mironoff, 42.
Monohybrids, 155.
Monsees, 72.
Morgan, 184, 194.
Mule hybrid, 250.

Mule hybrid — *Continued*
　and telegony, 170.
　mare secreting milk, 84.
Mutation theory, 154.
Mutilations, 207.

Nabours, 253.
Nathusius, 72.
Nature and nurture, 159.
Nucleoli, 6.
Nucleus, 5.
Number of young at a birth, 85.
Nutrition influences fertility, 98.

Œstrum, 41, 55.
　correlated with ovulation, 42.
　duration of, 63.
　effect of ration on, 64.
　influenced by lactation, 59.
　recurrence of, 63.
Origin of germ-cells, 46.
Ovaries, 24.
　removal of, 23.
Oviparous animals, 18.
Ovum, 13.
　fertilization of, 33.

Parturition, 75.
　mal-presentation, 77.
　normal, in domestic animals, 76, 78.
　treatment for mal-presentations, 79.
Pearl, 96, 97, 103, 108.
Pearson, 103.
Pedigree, 294.
Penycuik experiments, 168.
Poultry, age influences fertility, 94.
Practice of breeding, 280.
Pregnancy, early influence on mother, 274.
　heat during, 60.
　indications of, 66.
　physical examination for, 68.
Pregnant swine, high calcium rations for, 264.
Presence and absence hypothesis, 148.
Previous impregnation, influence of, 174.

Primary sexual characters, 19.
Prophase in cell division, 8.
Protein, effect of, on fœtus, 266.
Protoplasm, 1, 2, 4.
Puberty, 56.
    conditions influencing, 58.
Pure lines, 146, 291.
    Vilmorin's wheat-breeding, 292.

Ram, age influences fertility, 94.
Recurrence of œstrum, 63.
Reduction, 37.
    chromosomes, 37.
    in the female, 40.
    in the male, 41.
Registered breeding animals, 296.
Registry associations, 297.
Reproduction, 16, 183.
Reproductive organs, of the female, 24.
    of the male, 19.
Retarded growth, caused by premature breeding, 274.
    permanent effect of, 268.
Ritzema Bos, 232.
Rommel, 92.
Roux, 38, 39.

Secondary sexual characters, 19, 185.
Selecting the best, 300.
Selection, 284.
    aids to, 288.
    importance in animal-breeding, 286.
    methodical, 286.
    natural, 285.
Sex, 183.
    effect of age on determination of, 189.
    effect of nutrition on, 191.
    influenced by maturity of ovum, 192.
    not controlled by external conditions, 194.
    proportion influenced by season, 193.
Sex-linked characters, 188.
Sexual glands, transplantation of, 187.
    reproduction, 17.

Sheep, fertility influenced by age, 93.
Size, decreased by premature breeding, 272.
    improvement in, 280.
    of litter influenced by age, 91.
Somatoplasm, 161.
Sow, genital organs of, 30.
    size of litter influenced by age, 91.
Spallanzani, 43.
Spaying, 23.
Spencer, 71, 133, 167.
Spermatogenesis, 41.
Spermatozoa, 14.
    vitality within female generative organs, 46.
Spermatozoön in fertilization, 37.
Sperm-cells, conditions influencing vitality, 44.
    weakened by too frequent breeding, 45.
Steenbock, 64, 266.
Sterility, 113.
    abortion, 120, 123.
    causes, 114, 119.
    closure of cervix, 117.
    excessive food supply, 99.
    fatty degeneration, 120.
    in the female, 117.
    in the male, 114.
    obstruction in Fallopian tube, 118.
    twin births (free-martin), 129.
Stockard, 50.
Superfœtation, 60.
    examples of, 61.

Tanner, 104, 116.
Taylor, 126.
Telegony, 166.
    possible appearance in mule breeding, 170.
Telophase in cell division, 12.
Tessier, 70, 72.
Testicles, 20.
Theory of pure lines, 146.
Thomson, 133.
Thornton, 59.
Thury, 192.

Transmission of acquired characters, 157.
Transplanting sexual glands, 187.
Triplet calves, 108.
Trowbridge, P. F., 258, 268.
Twins, 101.

Union of egg and sperm, 36.
Unit characters, 141.
Use and disuse as causes of modifications, 212.
Uterus, 28.

Variation, 195.
    among cows, 203.
    causes, 210.
    continuous and discontinuous, 150, 151.
    examples of, 200.
    functional, 199.
    germinal, 214.
    infertility of animals, 200.
    meristic, 198.
    morphological, 196.
    physiological, 197.
    use and disuse, 212.
Verworn, 1.

Vilmorin's pure line wheat-breeding, 292.
Virchow, 2.
Viviparous animals, 18.
Von Guiata, 231.
Von Siebold, 192.

Waters, 257.
Weismann, 34, 38, 39, 167, 207, 210.
    and Von Guiata's in-breeding experiments, 231.
    theory of reduction, 38.
Wentworth, 81, 101, 107.
Wheat-breeding, 147, 292.
Wheat ration, 263.
    effect on fœtus, 263.
    effect of milk secretion, 267.
Wilsdorf, 230.
Wilson, 8, 33, 36, 38.
Wool production, improvement in, 282.
Wright, 74, 246.

Xenia, 176, 177.

Zebra hybrids, 168

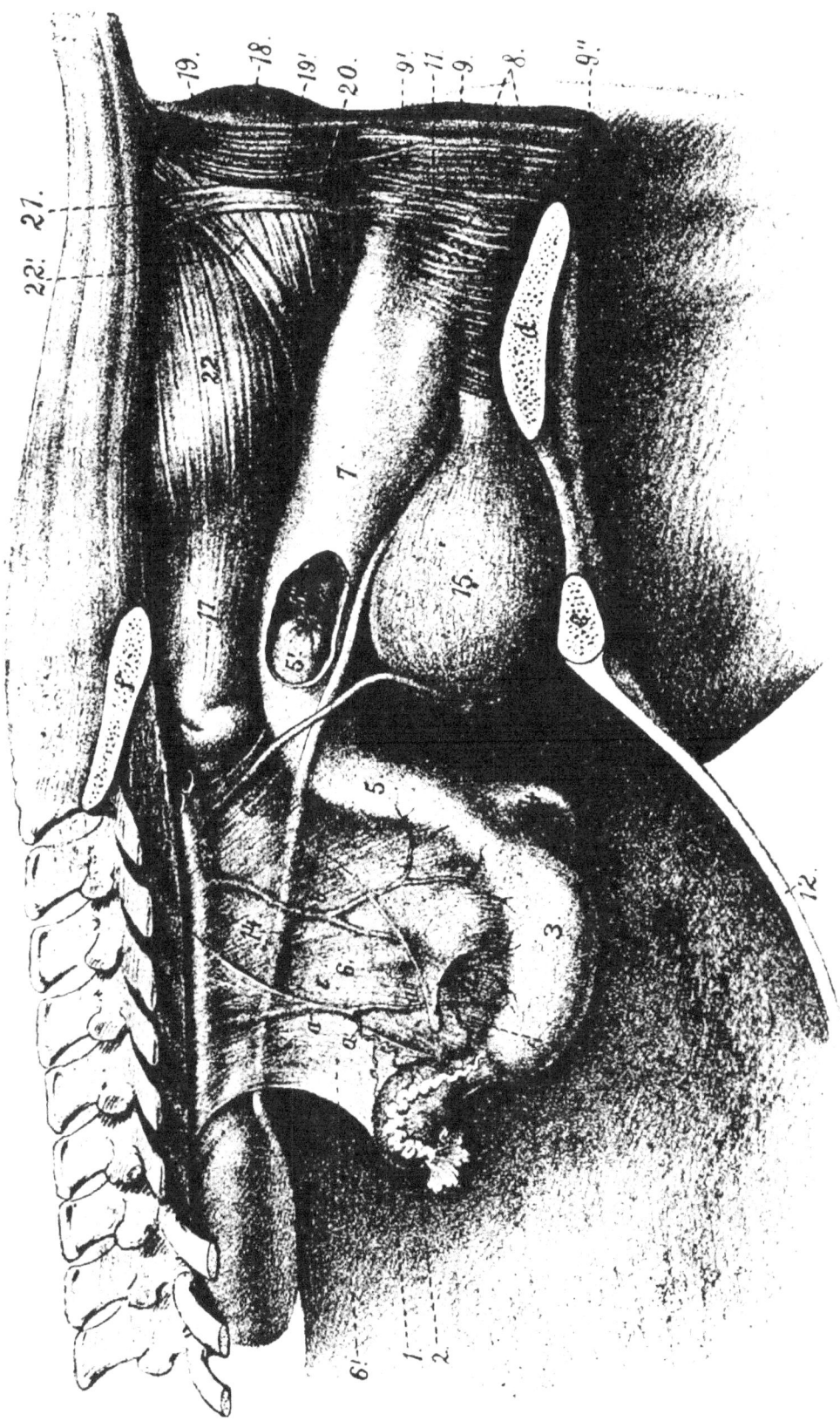

PLATE I.—Genital organs of the mare. In the non-pregnant mare the ovaries are in contact with the roof of the cavity. In this diagram these organs are shown pendant, due to the removal of the abdominal viscera. 1, ovary; 2, Fallopian tube; 3, left horn of the uterus; 4, right horn of the uterus; 5, body of uterus; 5' and 5'', cervix or neck of the womb; 7, vagina; 8, 9' and 9'', vulva; 10, constrictor muscle of vulva; 12, abdominal wall; 13, kidney; 14, ureter; 15, bladder; 16, urethra; 17, rectum; 18, anus. a, Utero-ovarian artery; a', ovarian branch; a'', uterine branch; b, uterine artery; c, umbilical artery; d, ischium; e, pubis; f, ilium.

PLATE XXXII. — Result of using a scrub sire. *Upper left hand*, a scrub ram; *upper right hand*, a grade western ewe; *lower*, a side and rear view of lamb from scrub ram, and western ewe. This lamb weighed fifty-six pounds at age four months, and sold for $4.50. Compare Plate XXXI.

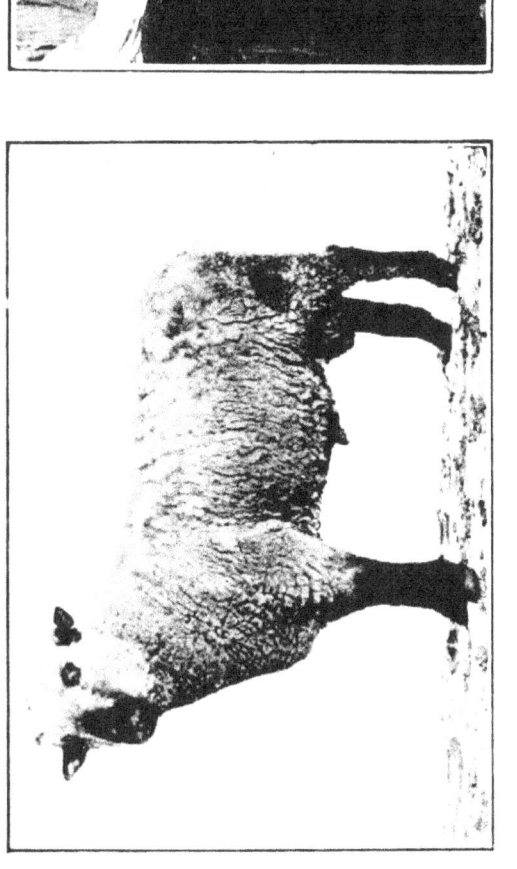

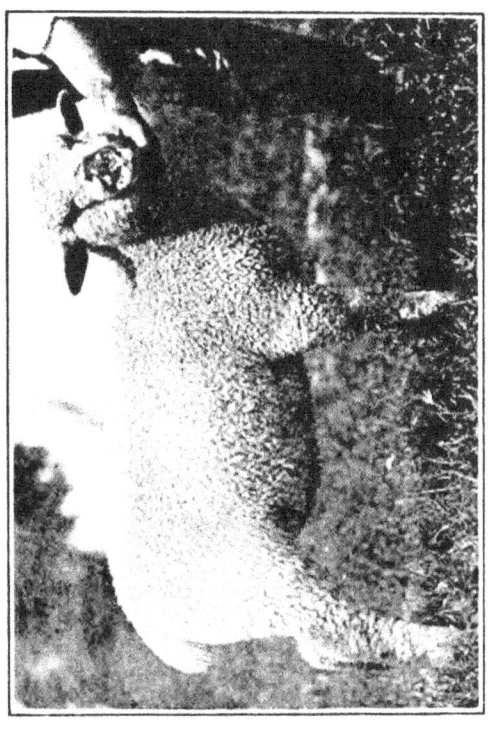

PLATE XXXI. — Result of using a pure-bred sire. *Upper left hand*, a pure-bred Hampshire ram; *upper right hand*, a grade western ewe; *lower*, a side and rear view of lamb from above parents. This lamb weighed sixty pounds at age three months, and sold for $7.35. Compare Plate XXXII.

PLATE XXX. — Three generations showing impressive character of original dam. *Upper*, Missouri Chief Josephine, yearly milk record 26,861 pounds. *Middle*, Missouri Josephine Sarcastic, daughter, yearly milk record

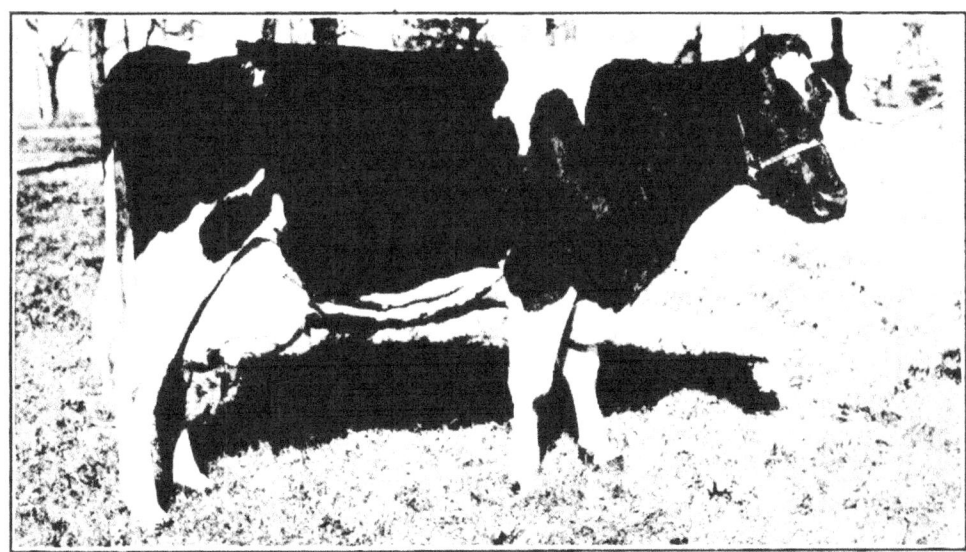
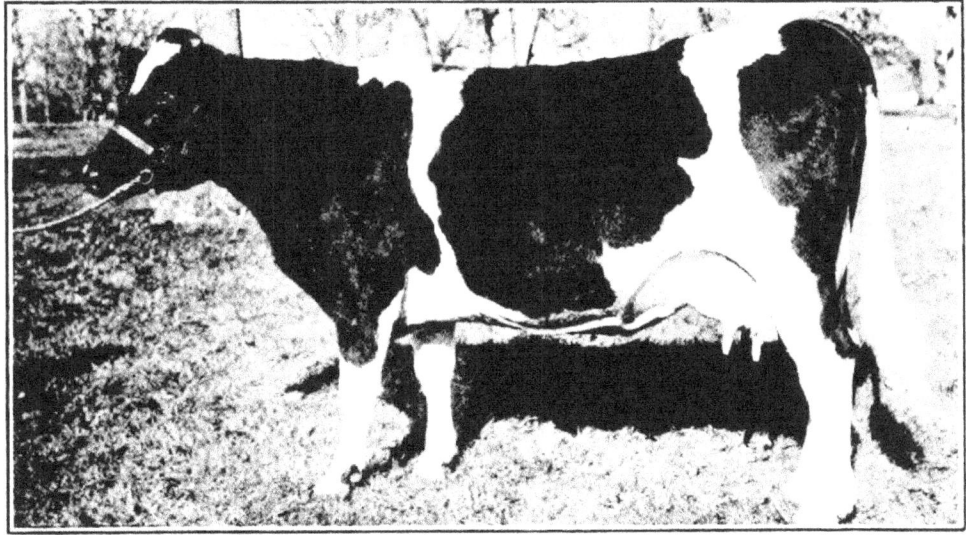

PLATE XXIX. — Daughters of the same sire, illustrating great uniformity in conformation and productive power resulting from the use of a prepotent sire in the herd. These three cows have an average yearly record of 22,295 pounds of milk and 855 pounds of butter. Owner, University of Missouri

PLATE XXVIII. — Result of selection by man. Sophie 19th, of Hood Farm; year's milk record, 1755 pounds, containing 999 pounds butter-fat.

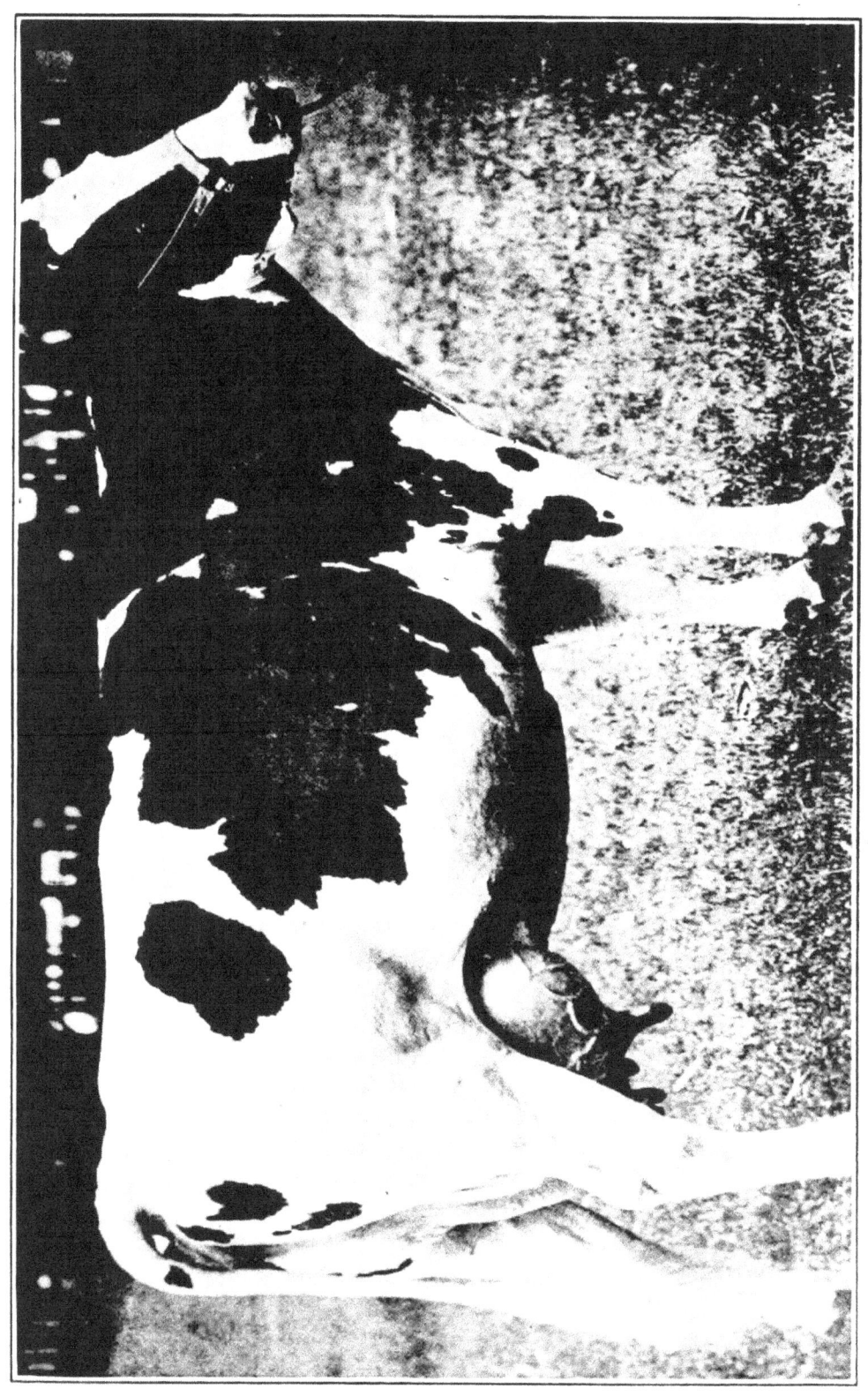

PLATE XXVII. — Result of selection by man. Duchess Skylark Ormsby, having produced 27,761 pounds of milk, containing 1205 pounds of butter-fat — the highest official record of any cow of any breed at this time (1916). Owner, John B. Irwin.

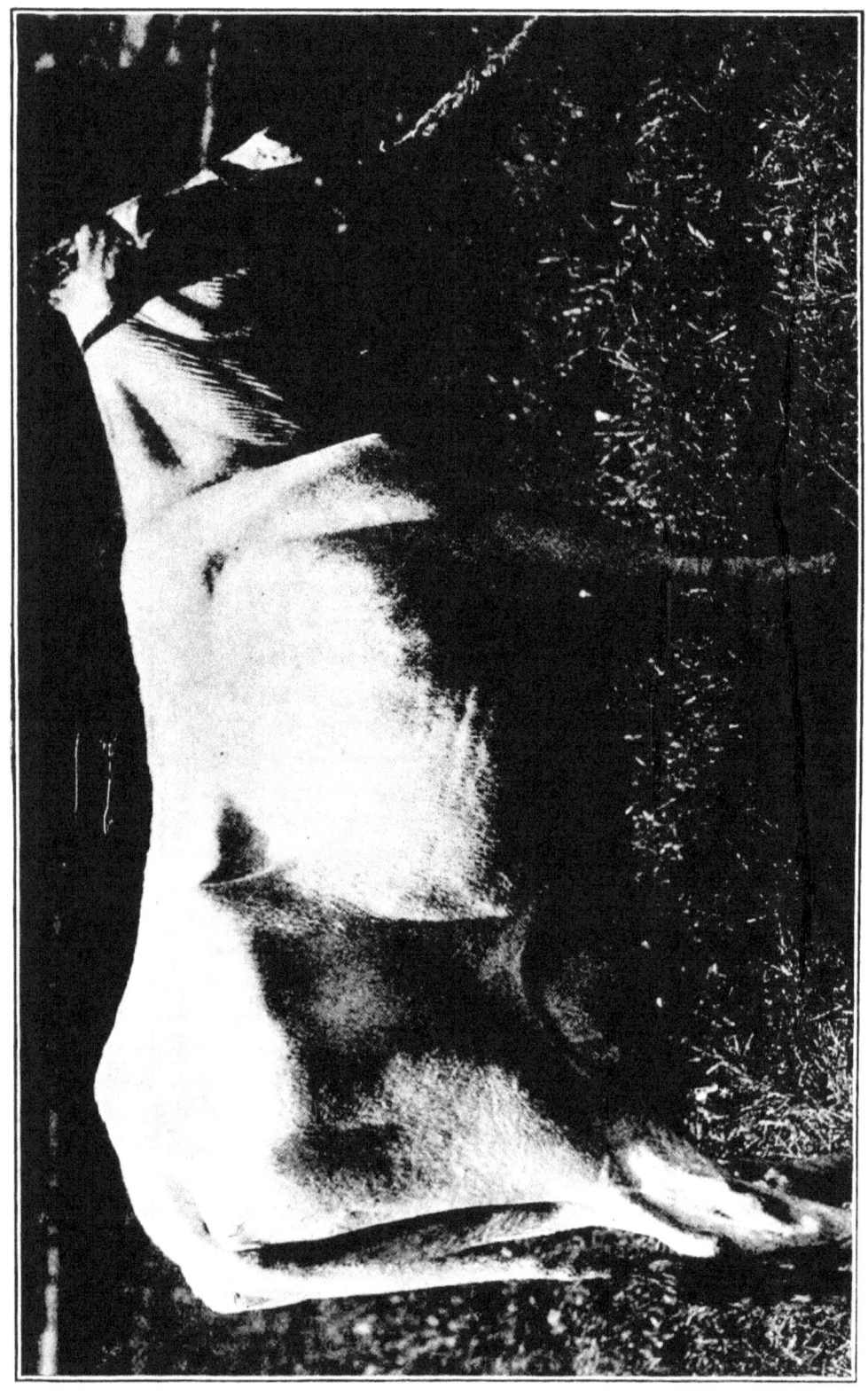

PLATE XXVI.— Grace Briggs at age 18 years. This remarkable cow has produced 111,309 pounds of milk and 5915 pounds of butter during her lifetime. Her best record for one year was 736 pounds of butter. In addition to this large yield she has produced 15 calves. Owner, University of Missouri.

PLATE XXV. — Permanent effect of retarded growth. The animal on the left (No. 527) was fed the maximum amount of a nutritious ration from birth. The animal on the right (No. 529) was starved until twelve months of age and then fed generously to age thirty-eight months. See table on page 269.

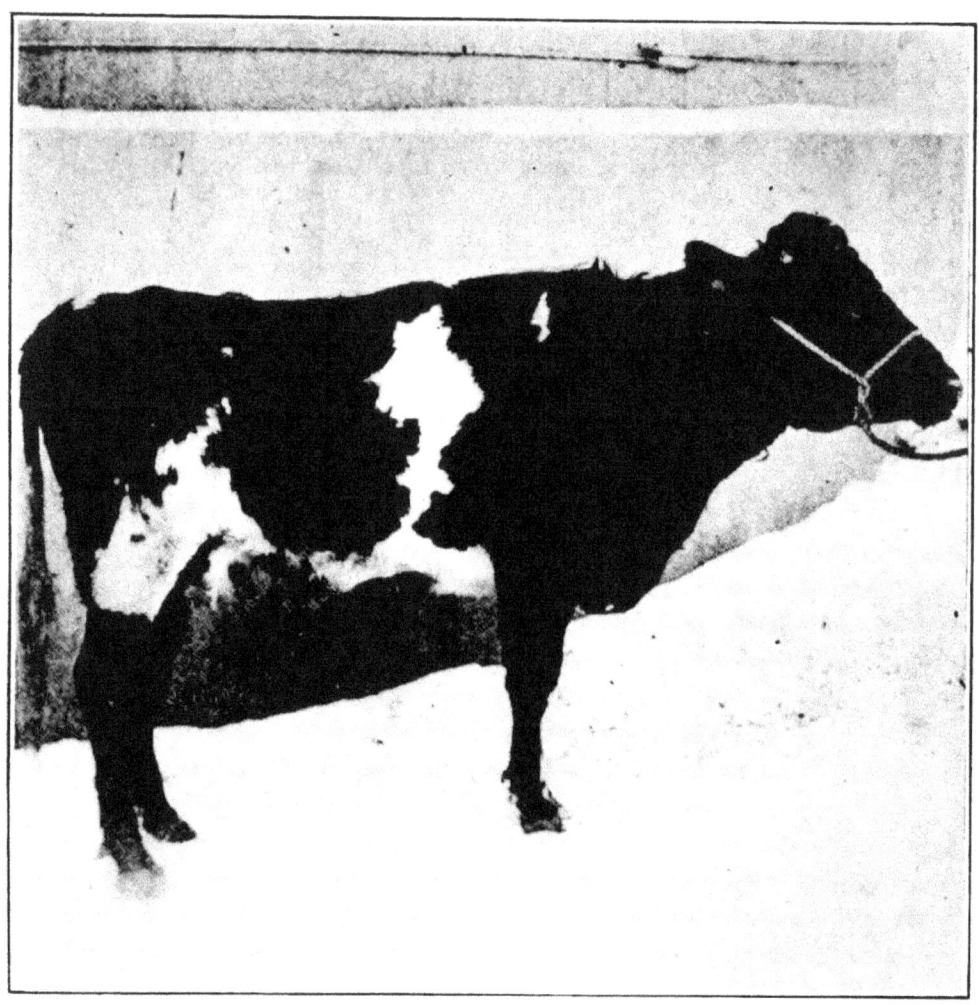

PLATE XXIV.—*Upper.* Unthrifty cow clearly in bad nutritive condition due to having been fed wheat products exclusively. *Lower.* Calf of low vitality from dam (upper) fed exclusively on wheat products, lived only twelve hours.

PLATE XXIII. — *Upper.* A normal healthy cow in good nutritive condition fed exclusively on nutrients derived from Indian corn. *Lower.* This vigorous, thrifty, well-developed new-born calf from dam (upper) fed exclusively corn ration before and during gestation.

PLATE XXII. — Result of starving. This animal was starved until twelve months of age, resulting in markedly retarding its normal growth. From age twelve months it was fed a full ration of nutritious food. Although greatly stunted by previous treatment, it grew and developed very rapidly, making unusually economical gains. Same animal as Plate XXI. Weight 1055 pounds at age 24 months.

PLATE XXI. — *Starvation does not destroy capacity to grow. Upper left.* Steer 529 weighing 175 pounds at age 96 days. Ration greatly restricted until age 365 days. *Upper right.* Steer 529 weighing only 200 pounds at age 310 days. Resulting from feeding greatly restricted ration. *Lower.* Same animal at age 38 months and weighing 1487 pounds.

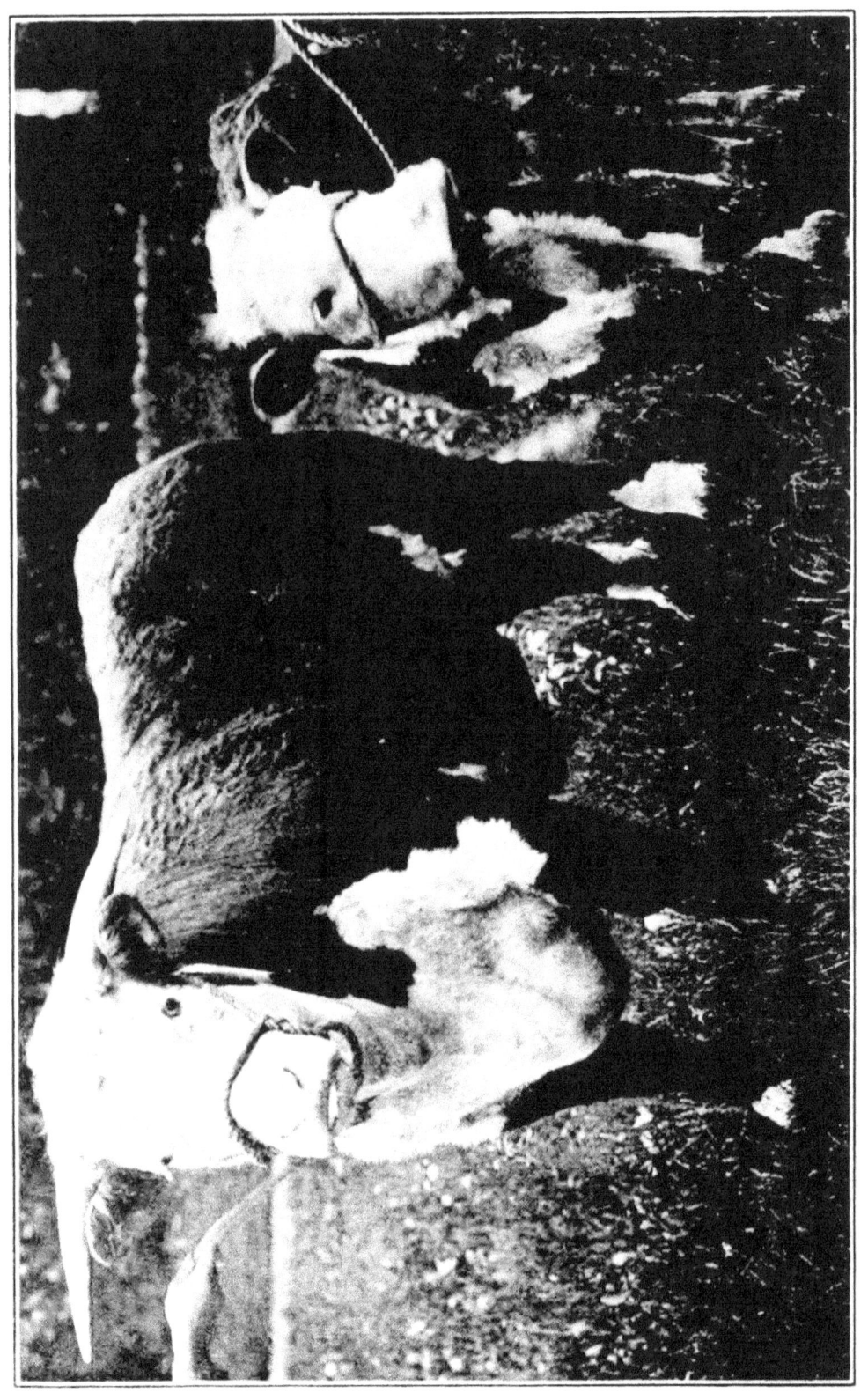

PLATE XX. — Effect of food on growth and development. Animals born on the same day photographed at age two years. Front view. See Plate XIX.

PLATE XIX. — Effect of food supply on development. These animals were born on the same day. The one on the left was fed generously from birth and weighed 1610 pounds when two years old. The one on the left was fed a ration limited in amount and at the same age weighed 361 pounds. See Plate XX.

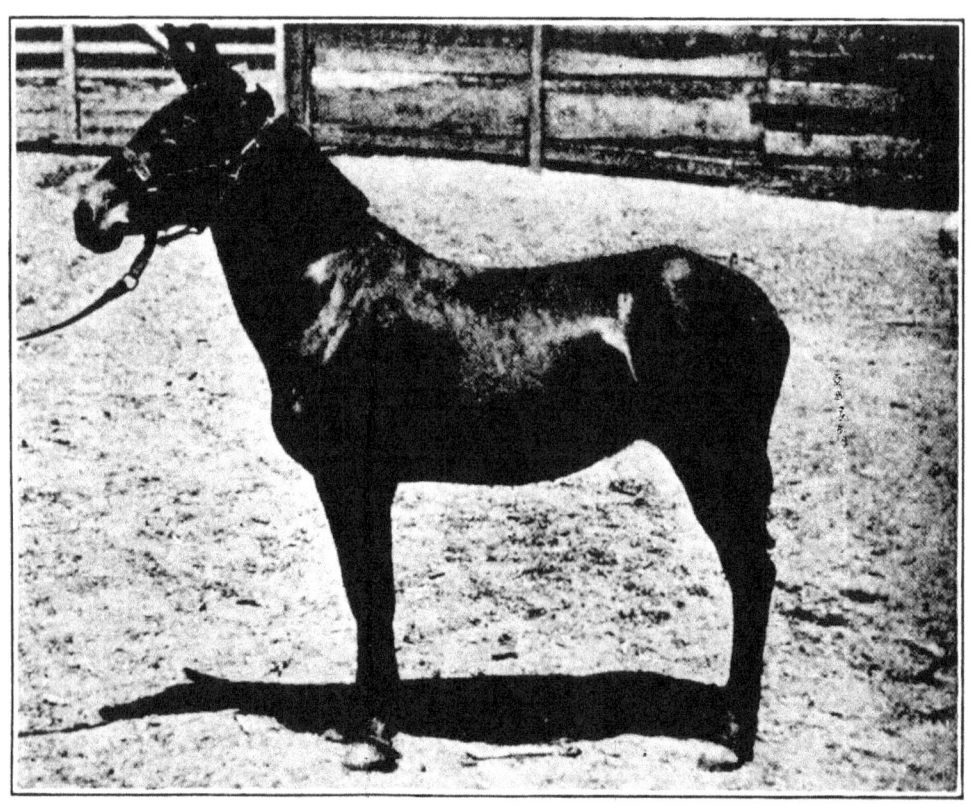

PLATE XVIII. — *Upper.* A five-year-old hinny. Dam a jennet, sire a Percheron stallion. *Lower.* Sheep-goat hybrid.

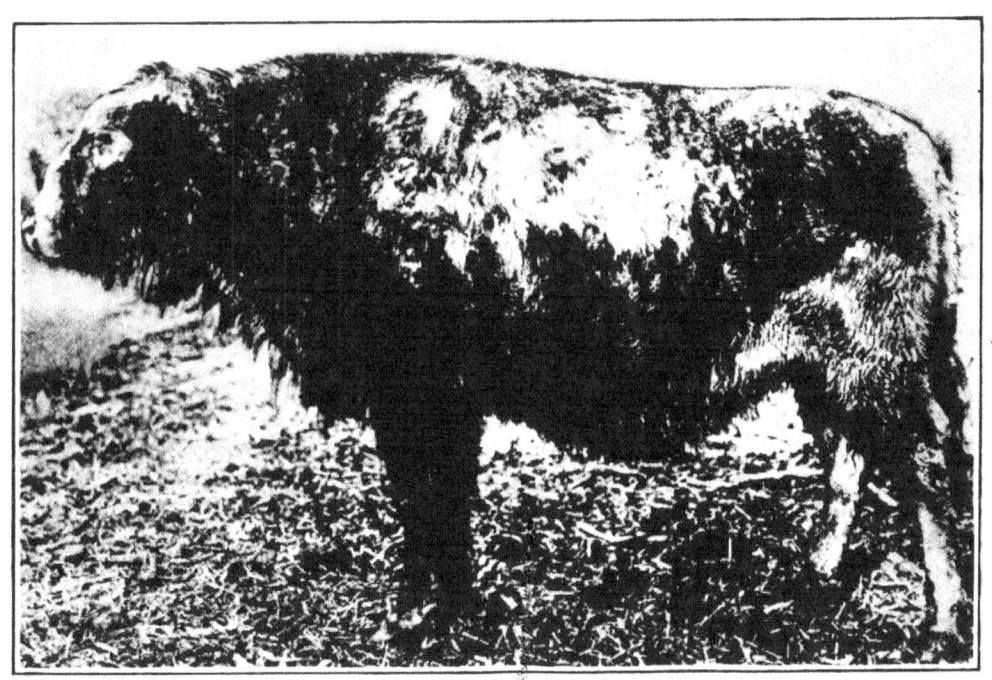

PLATE XVII. — *Upper.* — Half-blood buffalo (bison) heifer. The hybrids are larger than either parent. *Lower.* The bull on the left is five-eighths buffalo (bison) and three-eighths Hereford. The animal on the right is a three-quarter blood buffalo. These hybrids are frequently sterile.

PLATE XVI. — Deserter. A cross-bred Hereford-Aberdeen Angus steer. Champion at International Live Stock Show, 1909. Owner, University of Missouri.

PLATE XV. — In-bred Berkshire. Example of the good results which may follow a long line of continued in-breeding. Breeder, N. H. Gentry.

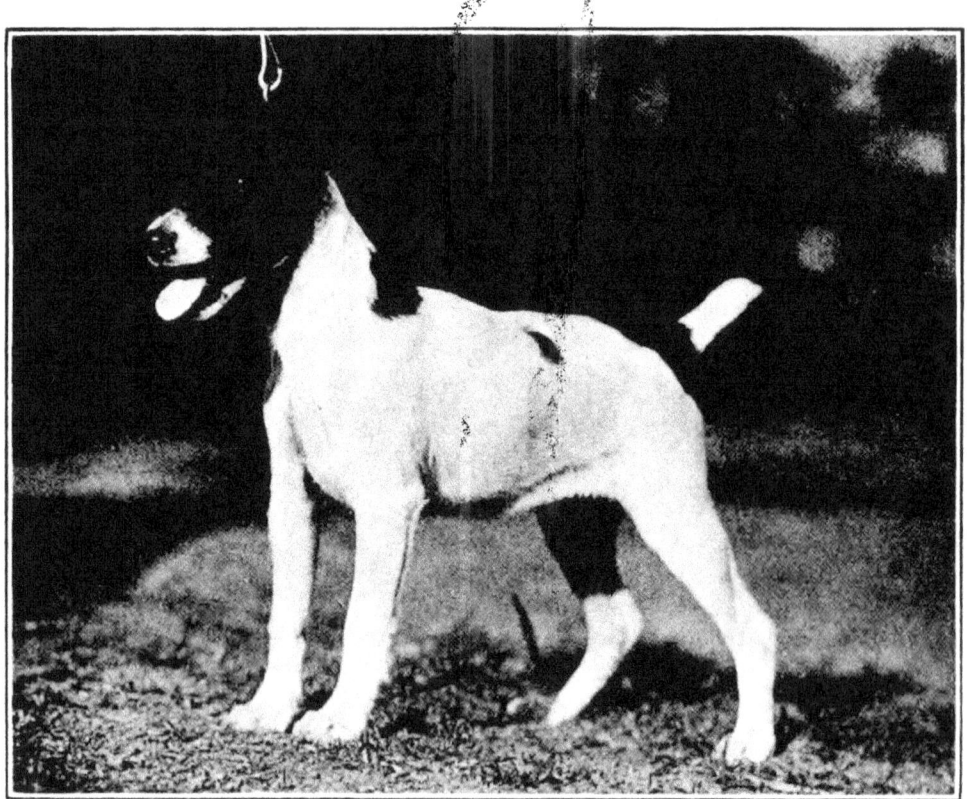

PLATE XIV. — *Upper.* Close in-breeding of fox terrier. "Dispatcher" at age nine months; ninth generation of intense in-breeding. *Lower.* "Designer 2d" at age four months; eighth generation of intense in-breeding.

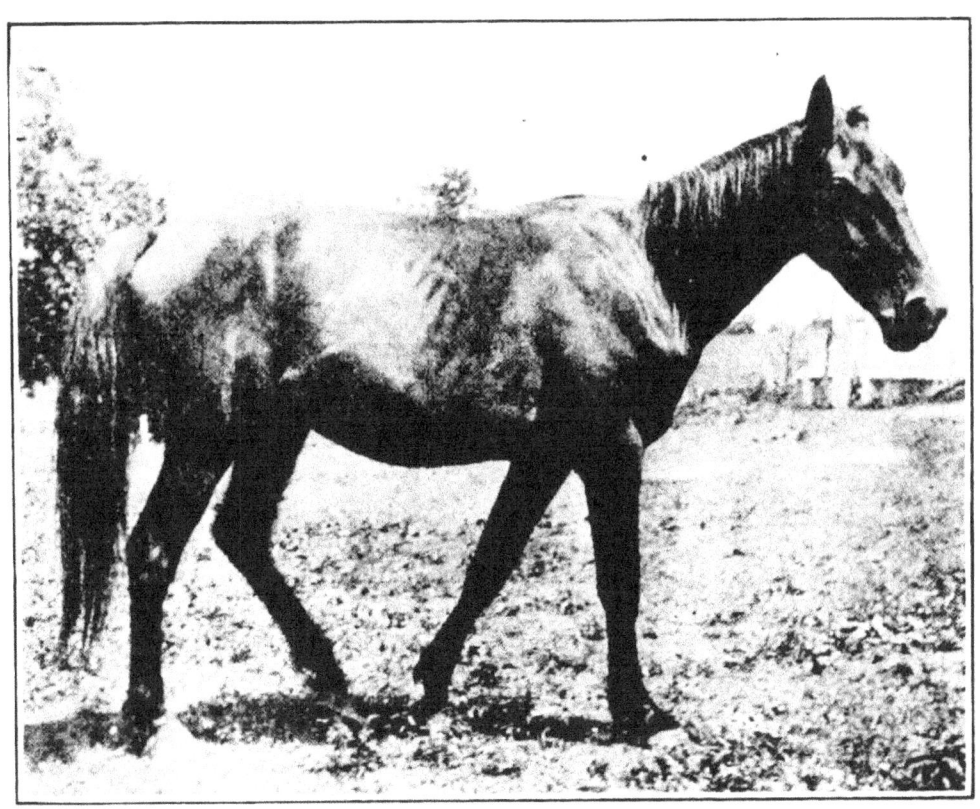

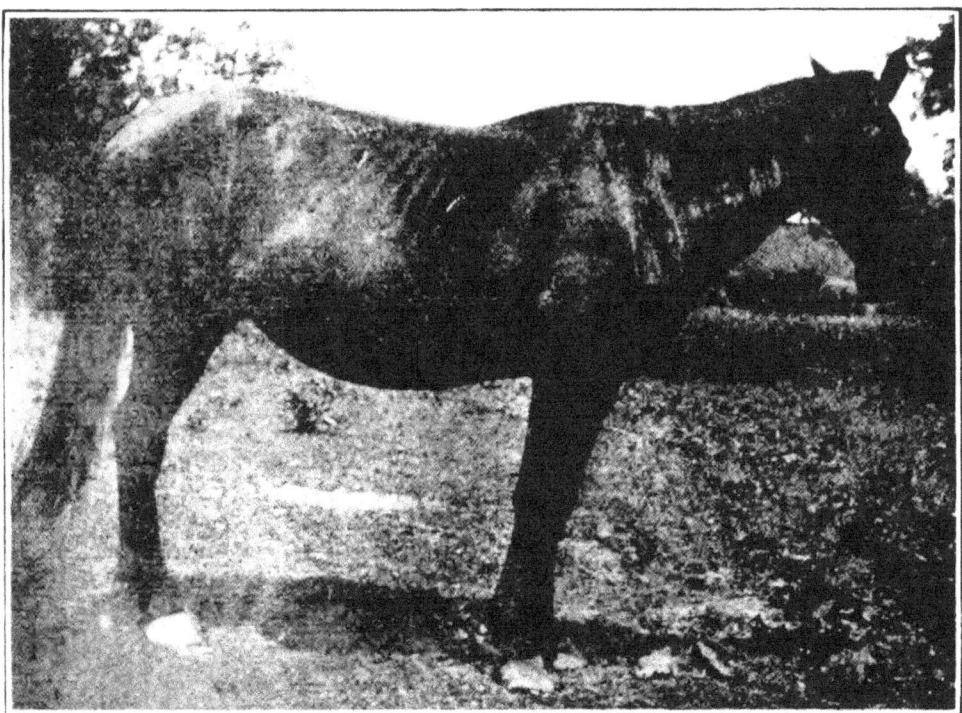

PLATE XIII. — *Upper*. "Crewdson." A condition favorable to the appearance of telegony. Previous to the birth of this mare her dam (Kate) had given birth to eleven mule foals in succession. *Lower*. "Kate," mother of eleven mule foals in succession, and later of "Crewdson."

PLATE XII. — *Upper.* "Sallie," the ninth foal following eight mule foals from same dam. *Lower.* The dam of this foal produced three horse foals, then eight mule foals and the animal shown in this illustration.

PLATE XI. — *Upper*. "Hallie," the eleventh foal of the dam "Maude" — the latter having previously given birth to ten mule foals in succession. *Lower*. "Maude," dam of "Hallie," produced ten mule foals previous to birth of "Hallie."

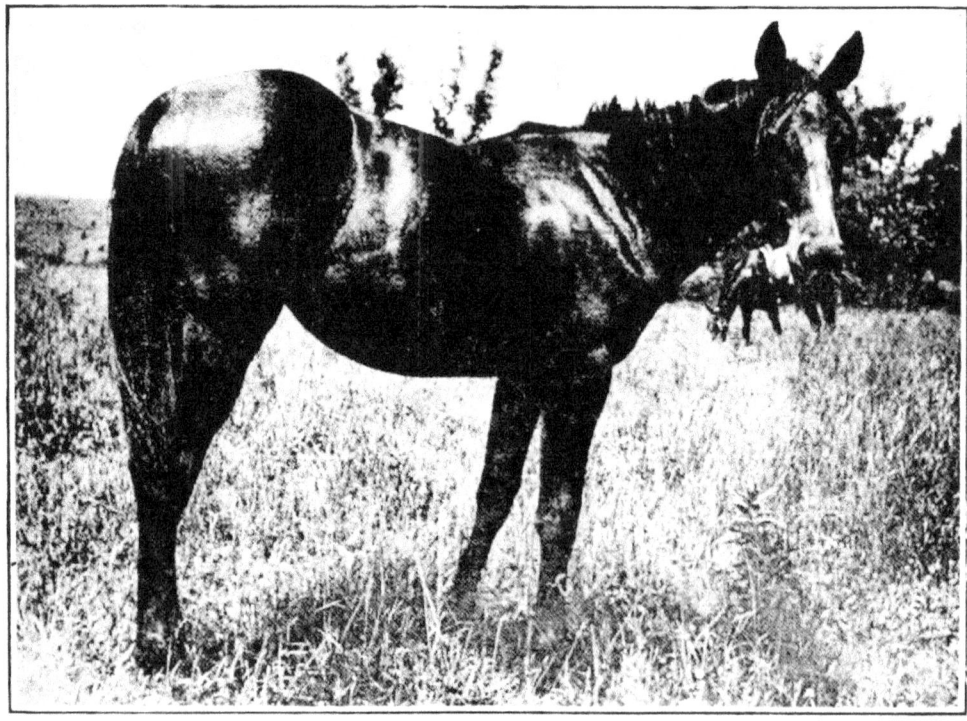

PLATE X. — *Upper.* Illustrating a condition favorable to the appearance of telegony in horses. This mare produced seven mule foals and then the mare shown below. *Lower.* The dam of this mare produced seven mule foals previous to the birth of this animal.

PLATE IX. — *Upper*. Steer 527 weighing 200 pounds at age 120 days. Fed maximum ration for rapid growth and development from age 90 days. *Lower*. Same, weighing 1453 pounds at age 27 months; rations not restricted.

PLATE VIII. — *Upper.* Ration not restricted. Fed for rapid growth and development from age 90 days. Age 336 days, weight 875 pounds. *Lower.* Ration partially restricted. Fed for normal but moderate growth and development. Age 780 days, weight 813 pounds, height 128 cm. Compare with Plate VII.

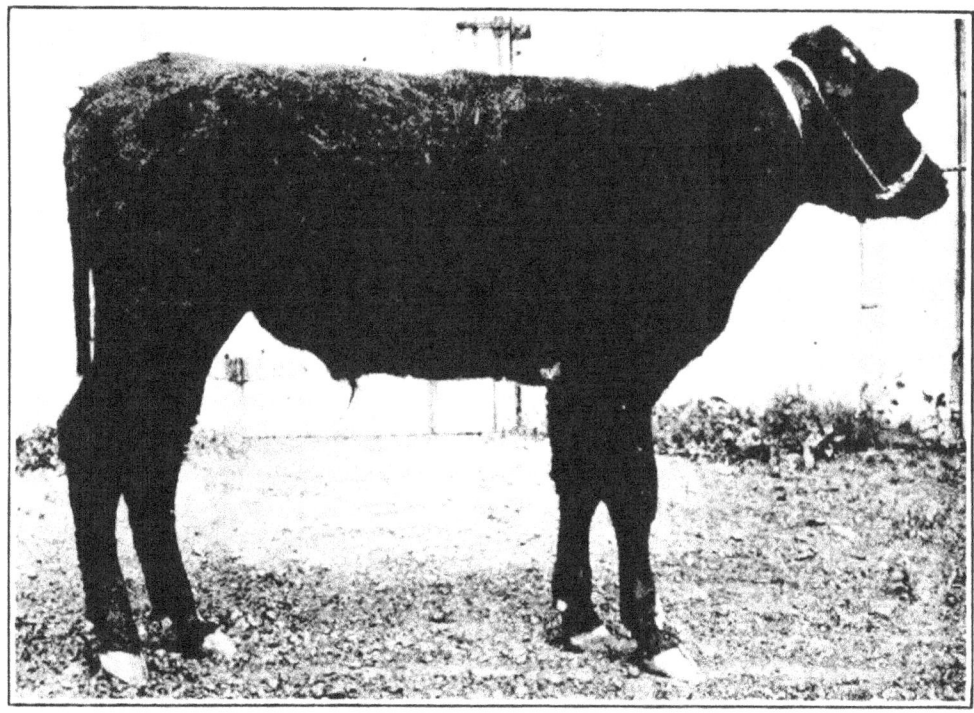

PLATE VII. — *Upper*. Ration not restricted. Fed for rapid growth and development from age 90 days. Age 716 days, weight 1401 pounds. *Lower*. Ration greatly restricted. Age 777 days, weight 512 pounds, height 121 cm.

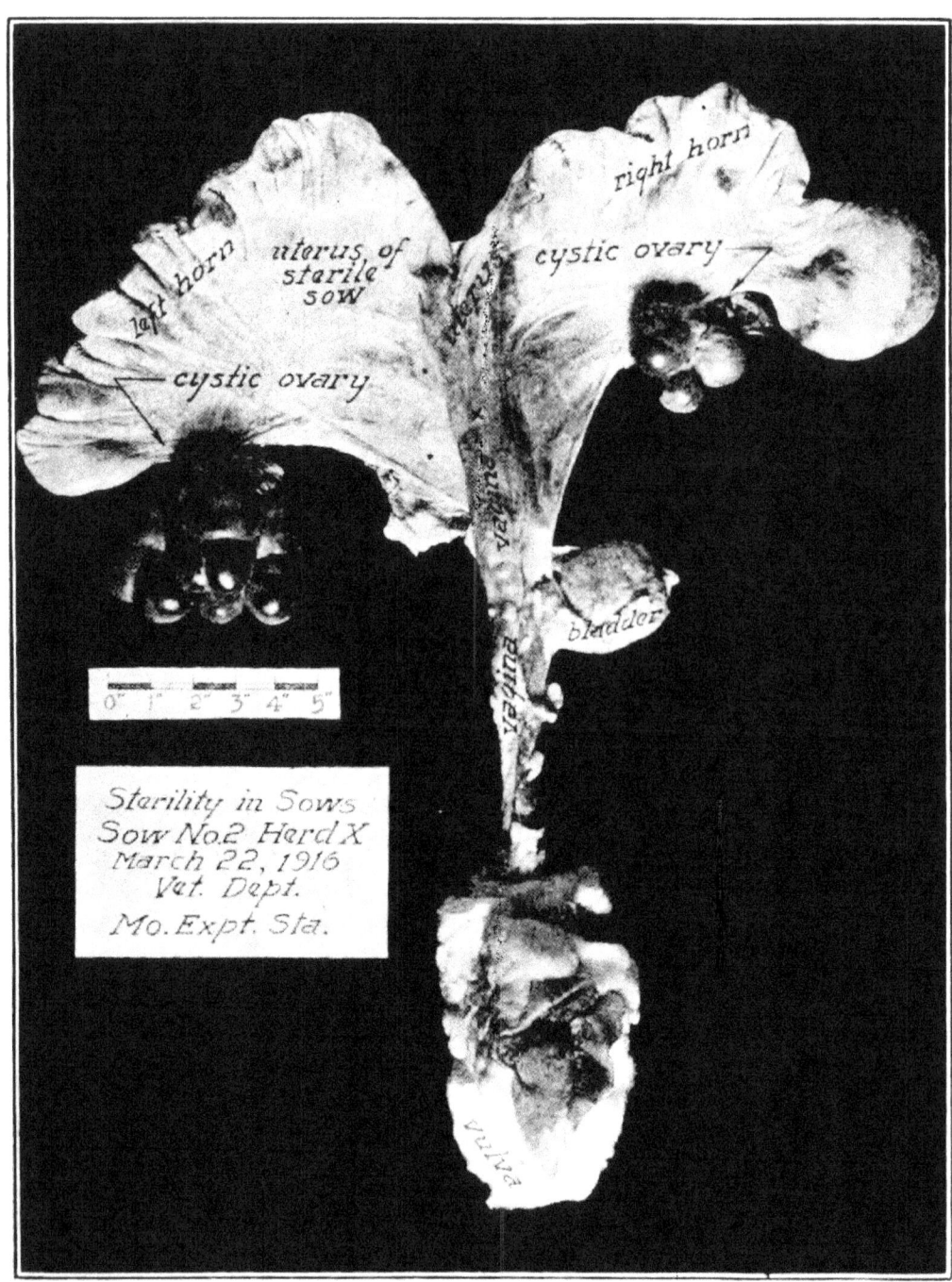

PLATE VI. — Sterility in the sow. Generative organs of a domestic sow, illustrating cystic ovaries which in this case resulted in complete sterility.

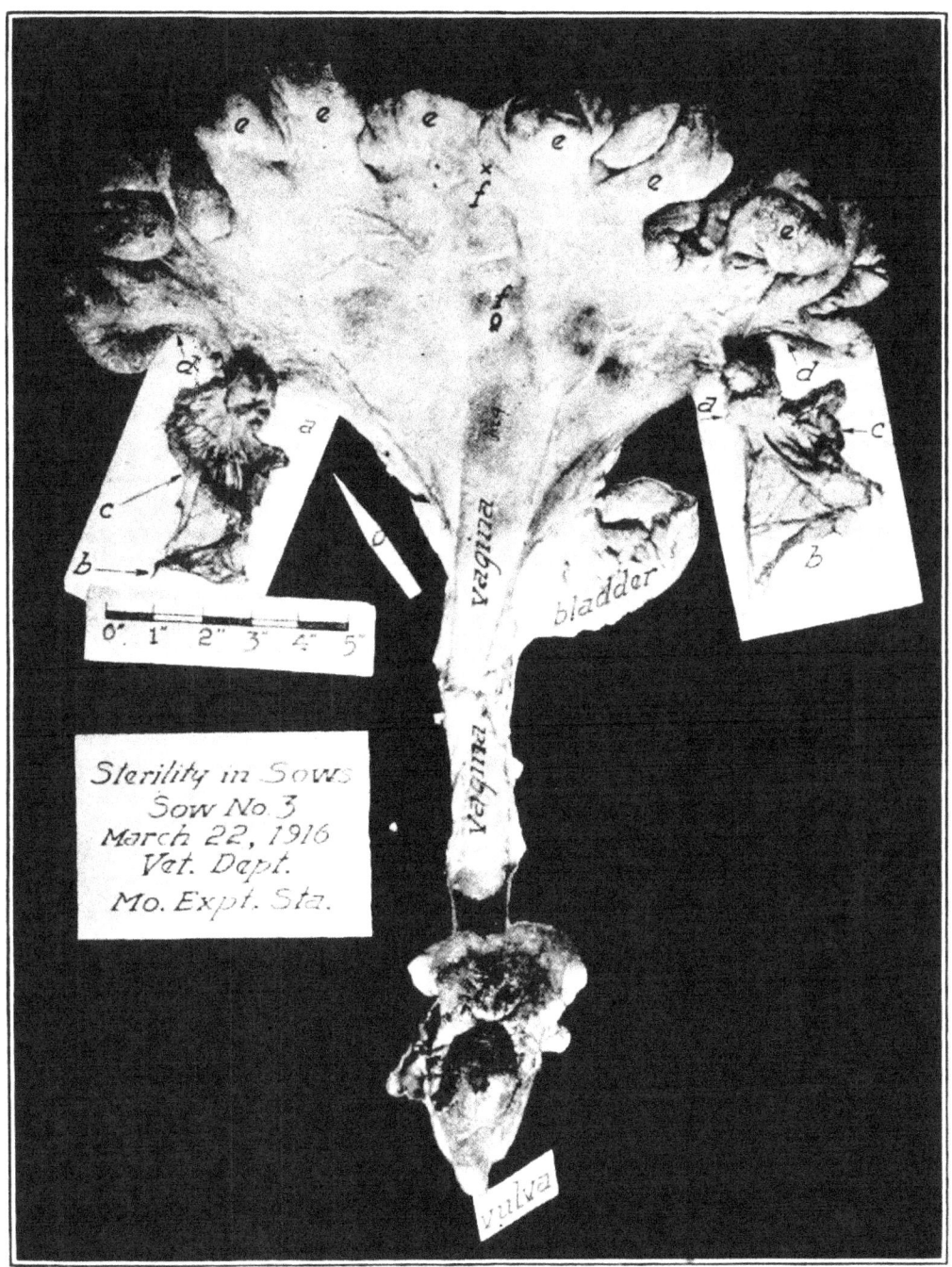

PLATE V. — Normal healthy uterus of sow. *a*, ovary; *b*, fimbriated end of Fallopian tube; *c*, central part of Fallopian tube; *d*, junction of Fallopian tube with horn of uterus; *e*, uterine folds of the horns of uterus; *f*, body of uterus; *o*, position of "os uteri"; *vag.*, vagina vulva, bladder.

PLATE IV. — Unusual fertility in the cow. Triplet calves from a grade Guernsey dam and grade Hereford sire.

PLATE III. — *Upper.* A mare mule that gives milk. *Lower.* A Free-Martin heifer that proved fertile.

PLATE II. — Superfœtation in mare. Twin foals, one a mule, the other a horse. Owner, A. L. Young.

www.ingramcontent.com/pod-product-compliance
Lightning Source LLC
Chambersburg PA
CBHW081140180526
45170CB00006B/1861